The Bee Researchers

Gudrun and Nikolaus Koeniger

Foreword by Hachiro Shimanuki

2026

Wicwas Press LLC
Kalamazoo, Michigan 49001 USA

The Bee Researchers

Copyright © 2026 by Gudrun and Nikolaus Koeniger

All rights reserved.

Published by Wicwas Press 1620 Miller Road Kalamazoo, MI 49001

www.wicwas.com

ISBN 978-1-878075-66-6

No part of this publication may be reproduced or transmitted in any form or by any other means, electronic or mechanical, including photocopy, Internet, recording or any informational storage and retrieval system known now or to be invented, without permission, in writing from the publisher, except for a reviewer who wishes to quote brief passages in connection with a review written for inclusion in a magazine, newspaper, broadcast or any form of Internet communication.

Dedication

We dedicate our book to the pioneers of the "Asian Honey Bee Universe":
Bruce A. Baptist, Roger A. Morse, Martin Lindauer,
Friedrich Ruttner, and Gordon F. Townsend

Acknowledgement

While much has been written about the post-war generation in West Germany, our experiences and memories do not fit into the general concept of "house builders" or "economic miracle profiteers". As bee researchers, we used the new freedoms of the post-war period to pursue our own personal path—to study the fascinating world of honey bees and conduct bee research.

The experiences described here were based on our personal memories. We represent many people from our point of view and our perspective of the respective events.

Without the substantial support and help we received, many things would not have succeeded. We met uncountable helpful people during our studies in Freiburg, Frankfurt, the bee Institute in Oberursel, Israel, Pakistan, Sri Lanka and finally in Canada.

Likewise, without the support of our friend and long-time companion Professor Robin Moritz, our initially German manuscript would probably not have been completed. Dr. Hachiro Shimanuki, a dear friend since 1969 and his wife Susan welcomed us to an unforgettable (wet!) evening in their beautiful home in Florida and encouraged us to produce this translation of the German book.

We owe the opportunity to transform the German "Die Bienenforscher" into "The Bee Researchers" to Larry Connor of Wicwas Press who tirelessly contributed to the creation of this beautiful book starting from our chaotic manuscript. We thank Hachiro Shimanuki and Richard Woodham for their editing talents!

Our sincere thanks to all!

Gudrun & Niko Koeniger   December 2025

Foreword

Husband and wives working together and independently in the same research area and still stay married are a rarity. Two famous such couples are the Marie and Louis Pasteur and Madam and Pierre Curie. "The Bee Researchers", Dr. Gudrun and Professor Nikolaus Koeniger are one such couple. Without doubt the most productive and famous bee research couple in the world, they both are well-known in their own speciality.

Gudrun is noted for her work on mating biology of the honey bees and Niko for his work on honey bees of Southeast Asia. Together and individually they have published hundreds of scientific papers and several books. "The Bee Researchers" is a love story of a woman who changes her career directions all for love of a man who loved bee research and discovers herself a love of bee research.

"The Bee Researchers" is a saga of their research, travels, and people that have contributed in the development of their careers. They share with the reader the difficulties of raising a family in a two-career family as told in their own words. Their story begins with both their early exposure to honey bees as children and by their incredible good fortune in being the only two students in the science curriculum that were allowed to begin their studies in the Fall of 1963. Thus began the start of their working cooperatively. The authors share their story of a hasty marriage, life in a war-torn Germany and the sociological after-effects of the Second World War. As beginning researchers and beekeepers they share their lessons learned from their mistakes. Included in their story are also their adventures of traveling on a motorcycle and anecdotes of people they encounter in their travels.

"The Bee Researchers" is a glimpse into the lives and research of the first 16 of 60 years of an ongoing research career of the Koenigers. As late as 2025 Gudrun and Niko participated in a exploratory trip to Malta to determine the presence of *Apis florea*.

—*Hachiro Shimanuki*

Table of Contents

Dedication	2
Acknowledgment	2
Foreword	3

Part 1   9
Bee Mecca on the Main and Research in Israel

First chapter   9
Niko finds Diogenes' barrel and a witch's hut with bees and Gudrun finds a biochemical wooden shack

Second chapter   12
Niko and Gudrun work in the bee Institute and ride the Horex motorcycle to Lunz in Austria

Third chapter   18
Gudrun and Niko catch drones in Lunz and help build a house on Lake Neusiedl

Fourth chapter   23
Gudrun and Niko start their doctoral theses and marry

Fifth chapter   32
Gudrun and Niko travel to Marseilles with the Horex

Sixth chapter   38
Gudrun and Niko dine French

Seventh chapter   42
Gudrun and Niko embark and meet Dwora and Fritz

Eighth chapter   46
Gudrun and Niko experience rejection, curiosity and also friendship

Ninth chapter   49
Gudrun and Niko come to Pardess Hanna, find the Ministry of Agriculture in Tel Aviv and reach Zrifin

Tenth chapter   54
Gudrun and Niko begin beekeeping and hear about the Displaced Persons after the war

Eleventh chapter   59
Gudrun and Niko don't want to be Dutch and meet Meir

Twelfth chapter   63
Gudrun and Niko are accused of espionage

Thirteenth chapter   68
Gudrun and Niko start their experiments

Fourteenth chapter   71
Niko gets an operation in the military hospital

Fifteenth chapter   75
Gudrun and Niko harvest honey, say goodbye to Horex and Israel

Part 2   78

Final sprint to the doctorate

Mother's happiness and father's career

Sixteenth Chapter   78

Gudrun and Niko settle back in Frankfurt and greet an aunt from America

Seventeenth chapter   81

Gudrun and Niko fight for bees for their experiments

Eighteenth chapter   84

Niko is visited by the MAD

Nineteenth chapter   90

Gudrun and Niko make a far-reaching decision

Twentieth chapter   96

Gudrun and Niko furnish a new apartment and submit their doctoral thesis

Twenty-first chapter   99

Gudrun and Niko complete their doctorates, become parents and do not accept a lucrative job offer

Twenty-second chapter   102

Gudrun and Niko stand on their own two feet financially and participate in the World Bee Congress

Twenty-third chapter   104

Niko fetches bee colonies in Pakistan

Twenty-fourth chapter   114

Gudrun and Niko have their second child, move and are happy to have visitors

Twenty-fifth chapter   116

Niko again fetches bees from Pakistan

Twenty-sixth chapter   127

Gudrun is involved in the "Kinderschule" and in the local SPD (Social democratic Party Germany)

Twenty-seventh chapter   132

Gudrun finds her own way

Twenty-eighth chapter   136

Gudrun returns to bee research

Twenty-ninth chapter   138

Niko fetches Stefan from Darmstadt

Thirtieth chapter   141

Niko and Stefan hiss at bears

Thirty-first chapter   143

Gudrun and Niko attend a congress in London and Niko becomes a lecturer

Thirty-second chapter   146

Niko changes his travel plans and Gudrun arrives in Sri Lanka with the children

Thirty-third chapter   150

Niko searches for the giant honey bees and Gudrun settles in Kandy

Thirty-fourth chapter　　　　152

Niko finds giant honey bees and Gudrun is visited by the police

Thirty-fifth chapter　　　　159

Niko works on Asian giant honey bees in Oberursel and plans another trip to Sri Lanka

Thirty-sixth chapter　　　　163

Gudrun and Niko arrive in Anuradhapura and rent a house with a watchman

Thirty-sixth chapter　　　　169

Gudrun and Niko celebrate Christmas under a palm tree

Thirty-seventh chapter　　　　175

Niko and Noel observe drone flight

Thirty-eighth chapter　　　　179

Gudrun and the students offer Lego bricks to the bees

Thirty-ninth chapter　　　　183

Gudrun and Niko have a visitor

Fortieth chapter　　　　188

Gudrun and Niko drive to the sea and catch the return flight at the last minute

Forty-first chapter　　　　191

Niko works in Hohenheim and applies for a bee Professorship in Canada

Forty-second chapter　　　　194

Gudrun and Niko visit Guelph in Canada

Forty-third chapter　　　　198

Gudrun and Niko clear up misunderstandings

Forty-fourth chapter　　　　201

Gudrun, Niko and the children emigrate to Canada

Epilogue 2025　　　　206

Prologue

Gudrun, a girl born two years before the end of Second World War, is interested in everything that crawls or flies. In front of her grandfather's apiary in the northern town of Wilhelmshaven, she asks: "How do the bees always find their way home?" "Oh, that's just the way it is. They do it instinctively," he answers. A bad answer, the girl thinks. "Instinctivey" means that the grandfather knows nothing about it. Gudrun likes her grandfather and the bees. She swallows her disappointment.

Later, Gudrun reads the books of Niko Tinbergen and Konrad Lorenz. "I want to be an animal psychologist!" she says enthusiastically to her father, who heads a scientific Institute as a geologist and prehistorian. "I want to understand the behavior of the animals! Of dogs, chickens and of honey bees."

Her father is firmly against it: "Dear daughter, being an animal psychologist is not a profession for you! Bee research in particular has no future. As a woman, you will marry and raise children. If it has to be a science subject, you should become a teacher. Then you have time for your family."

Gudrun finished her Gymnasium with Abitur (final examination) in the spring of 1962. Since she was needed at home to care for her seriously ill grandmother and the child of her sister, she delayed her education. When her sister returned in October 1962 she enrolled in the University of Freiburg in the south of Germany, most distant from her home in Wilhelmshven. Normally the biology curriculum in German universities start in March however she was allowed to become one of two first semester students in the autumn of 1962.

Niko, a boy born four years before the end of Second World War, is interested in everything that crawls or flies. Once again, he is standing in front of the apiary in Bad Salzuflen with his father, a doctor. The bees fly busily in and out. "How do the bees always find their way home?" he asks his father. "Oh, that's just the way it is. They do it instinctively," the father answers. "Instinctively! That's always your excuse if you don't know. I want to know more, find out more. I'm going to be a bee researcher!"

His father is firmly against it: "Dear son, bee researcher is not a profession for you. This is an exotic subject without a future. As a man, you will have to feed your wife and children later. Niko, you should study medicine. As a doctor, like me, you can keep some bee colonies. Bees are a nice hobby, not a profession!"

Niko finished his Gymnasium with Abitur (final examination) in the spring of 1961. He then joined the Bundeswehr, compulsory military service which lasted until the summer of 1962. Then in autumn 1962 he enrolled in the biology curriculum of the University of Freiburg in the south of Germany most distant from Bad Salzuflen. In the zoological internship of the University he became the second of the two first semester students, the first being Gudrun. The two students were placed side by side. Their first contacts focused on shared and deep interest in the sciences, biology, zoology and especially honey bees. Beyond their studies, this cooperation expanded to regular joint undertakings and activities. At the end of five semesters at the University of Freiburg the relationship of Gudrun and Niko grew to a partnership based on love, and cooperation.

On March 1965 a large motor scooter, a "Heinkel Tourist," coming from Freiburg, turns onto the motorway north in the direction of Karlsruhe and Frankfurt. Very gradually the heavily loaded scooter accelerates to its final speed of about 75 km/h. It is cold. The sun is shining, the road is dry and both passengers listen to the sonorous and even soothing sound of the motor.

The pillion rider, Gudrun, shivers. Her back though covered by a large rucksack that reaches from her shoulders down to the back seat becomes gradually colder. In front she clings to the back of the driver: Niko's back transfers comforting heat.

"Yes, here I am going to Frankfurt with Niko. I would have preferred to stay in beautiful and idyllic Freiburg. Niko wants to get into honey bees and he is convinced that the University of Frankfurt will turn into a "Mecca" of honey bee science. Frankfurt, that's where Niko will go!" Gudrun thinks. Memories of her beloved grandfather and his bee hives become alive! The high academic reputation of the biochemistry department at Frankfurt University is also important to her. "I want to keep the focus of my studies in biochemistry," Gudrun's thoughts continue: "The decisive factor for my move to Frankfurt is ultimately the partnership with Niko, which is dear and important to me."

The cold wind blows into Niko' face. His back, however, is comfortably warm! Gudrun on the back seat has snuggled up close to him, probably not only to avoid the wind. "Yes, this snuggling has become a perfect routine with Gudrun", he reflects. "It didn't start easily: Exaggerated dogmatism on my part deterred Gudrun. Further my concentration on zoology and the initial rejection of non-zoological subjects led to conflicts. The joint course works, lectures and exams then resulted in a professional, successful cooperation with Gudrun. Later, many excursions and personal undertakings with Gudrun became an important part of daily life." Niko is not sure whether or not he would have left Freiburg without Gudrun! At the end, Gudrun's decision after long discussions to come to Frankfurt was more than welcome! "The partnership with Gudrun is dear and important to me," is Niko's conclusion.

Part 1

Bee Mecca on the Main and Research in Israel

First chapter
Niko finds Diogenes' barrel and a witch's hut with bees and Gudrun finds a biochemical wooden shack

Early in March 1965, Gudrun and Niko arrive in Frankfurt. They find temporary shelter with Niko's Aunt Anna in "Holbeinstrasse in Sachsenhausen," a southern quarter of Frankfurt. With mixed feelings—the farewell to Freiburg is not yet over—they both take a walk to the city center the next morning to get to know their future home. Niko's aunt recommended the path along the banks of the Main river, over the "Eiserner Steg," to Frankfurt Cathedral and to the "Zeil," a road with countless shops. It is frosty, the sun is shining and the banks of the Main invite you to linger. They find a free bench on the waterfront, let the spring sun warm them up and watch the ducks and black-headed gulls bobbing peacefully on the Main river.

Over a coffee on the other side of the Main, they discuss the whole thing. Frankfurt actually has little in common with the student's idyll of Freiburg. Ruins are still standing everywhere, remnants of the hail of allied bombs 20 years ago. New buildings dominate the city centre. However, the area in the heart of the city between the "Römer" and the "Kaiser Cathedral" is still a field of rubble.

Aunt Anna subscribes to the Frankfurter Rundschau, a daily newspaper with a large advertising section listing rooms and apartments. Gudrun comes across a room in the "Römerstadt" with good connections with tram and buses. The rent—110 marks—is reasonable. Niko is less lucky. What he finds is either too expensive, too far out or has other restrictions! Finally, he places an advertisement himself. His ad:

"I feel comfortable even in a barrel! Student is looking for a room" brings immediate success! There is a flood of mostly funny answers that underline the Greek classical background.

"We are pleased Diogenes, that you finally arrived in Frankfurt!"

and

"Diogenes in his barrel, Doesn't like to quarrel He would like to find a room, Not later, but better soon!"

In addition to the satire, there were also two serious offers: A large discarded railway-carriage in a garden in Ginnheim. Quite to Niko's taste but during inspection it becomes clear that without heating it would only function as a summer accommodation. In

"Kleine Seestraße", the center of Bockenheim there was a room that can only be accessed via a ladder: twelve square meters and a ceiling height between 1.80 and 1.40 meters for 60 marks per month. In the anteroom a small sink. Toilet at the bottom next to the ladder. The small window above the Esso petrol station allows a view of Bockenheim: "Große Seestraße and Mühlgasse". The Zoological Institute is within walking distance. Top location! Niko moves in immediately.

The house belongs to the Werner family of craftsmen, electricians. Niko's question as to how such a room can be created quickly finds an explanation: Bockenheim was largely destroyed by bombing raids. Three outer walls of the Werners' house had remained upright. As soon as the first sack of cement could be procured, walls were erected with bricks from the surrounding rubble and the first urgently needed living space was created. So it went on step by step, without an architect and without official building supervision, until Niko's new home on the second floor, above the electrical workshop, was completed.

Fig.1. Diogenes barrel: "Kleine Seestraße". Niko's window with flower box!

The Institutes of Zoology and Botany are located together inside a green area of the city, nestled between the "Palm Garden," the "Grüneburg Park" and the "Botanical Garden." There were functional concrete buildings from the post-war period, as in Freiburg. All biology students enter the palm garden free of charge. This shortens and beautifies the daily walk to the student's canteen at the Bockenheimer Warte. During the lectures and internships in Frankfurt, Gudrun and Niko experience many repetitions of subjects they had learned in Freiburg. The lectures and practical experience of Professor Martin Lindauer impress them. His enthusiasm for biology keeps all students under his spell—and bees are almost always the center of attention.

Right at the beginning, Gudrun and Niko call on Dr. Jander, whom they met in Freiburg. He now greets them happily and immediately hires them for further experimental work. It is about the visual shrub-specific recognition system of stick insects. These studies are to be extended to various animal groups, including honey bees. This job, together with the support of their parents, secures an adequate monthly income in Frankfurt.

Niko wants to learn as much as possible about the Bee Institute in Oberursel. In the library of the Zoological Institute, he finds a few detailed documents on the subject, and locates a "contemporary witness." the entomologist Professor Peter Rietschel: "The Bee

Institute was founded in 1937 from an investigation center for bee diseases," says Rietschel. "In response to large losses of bee colonies in Germany, it was founded first at the Zoological Institute in Frankfurt. Later the "Polytechnische Gesellschaft" (Polytechnic Society), a citizens' foundation of Frankfurt, has taken over the Institute and settled it in Oberursel. After the war, the teacher Hugo Gontarski became the director of the Bee Institute. Self-taught, he did excellent research, both in the fight against bee diseases and in honey analysis. His quality test for honey is still considered an international standard to this day!"

Niko has read several interesting articles by Hugo Gontarski in beekeeping magazines.

"Nevertheless, many colleagues within honey bee science did not recognize him. Envy and resentment made life difficult for poor Gontarski," says Professor Rietschel. "Just because he didn't have a doctorate?" Niko shakes his head: "Science is science, with or without an academic title."

"It was only when Hugo Gontarski died unexpectedly in 1963 that the Polytechnic Society suggested to Professor Lindauer that the Bee Institute be re-attached to the University of Frankfurt and continued as part of the zoological Institute. This offer was accepted by the university and the financing, the material and personnel equipment for the bee Institute, was contractually stipulated."

Niko understands: That's why Dr. Friedrich Ruttner was appointed professor of zoology at the University of Frankfurt and at the same time became head of the Bee Institute in Oberursel.

Niko is keen to see the Bee Institute. There is a tram that leads to Oberursel. Niko, however, rides on his scooter, the Heinkel. The roads lead north, first through suburbs, then past fields, always along the tram rails until Oberursel begins, a small town at the foot of the Taunus mountains. The journey continues through the town, past an engine factory, on to the US Army base "Camp King", until just before the terminus of the tram. Niko had not imagined it to be so rural. He finds the street "Im Rosengärtchen" and at the end of the cul-de-sac, right at the edge of the forest, there is finally a small house. This can't be an Institute of the university, can it? Niko drives back to see if he has made a mistake.

Fig. 2. Institute of Bee Research in Oberursel (University of Frankfurt). 1965.

No, everything seems to be ok. And then he also sees an apiary with colorful beehives. Everything is very narrow and small. Compared to the Institutes at the university, the building looks old and quite neglected. A little paint and some new roof shingles would be appropriate. This is where Hugo Gontarski is said to have developed his famous enzymatic honey analyses? Inconceivable! Can modern research function under such dilapidated and cramped conditions? In any case, there is no sign of a "Mecca" for honey bee science here. Did Gudrun and he leave Freiburg because of a castle in the air? Because of this shack? Disappointed and insecure, Niko returns from Oberursel. However, he describes his impressions positively to Gudrun:

"It is often small research institutions that are very successful. Professor Ruttner is creative. Together with Lindauer, he will succeed in expanding the Institute in Oberursel."

Meanwhile, Gudrun has learned about the Biochemistry Institute—and is also rather disappointed with the working conditions.

"Hopefully you're right about the small successful research units. The chemical Institute is housed in an old brick building on "Robert-Mayer-Straße." People sit on the third floor, all in a single room crammed with chemical equipment and the professor sits behind a wooden shed. If the devices were a little heavier, everything would probably collapse." She shakes her head: "Nevertheless, Pfleiderer's working group has many publications in very good journals! The professor took time for me and I am welcome to work with them during the semester break—but without pay!"

Gudrun and Niko look at each other and hope that the decision for Frankfurt was the right one.

Second chapter

Niko and Gudrun work in the bee Institute and ride the Horex motorcycle to Lunz in Austria

Niko has persuaded Gudrun to listen to Professor Ruttner's first lecture. In the small lecture hall of the zoology department, Professor Ruttner talks about the biology of the honey bee. He walks hunched over, pulls one leg and holds his left hand in a cramped position. Are these the consequences of a stroke, Gudrun and Niko ask themselves worriedly? Professor Ruttner drags himself clumsily to the lectern. His hair is thinning and overall, he gives the impression of an old man. Can someone like that do dedicated research? Isn't the disability too great a handicap?

As soon as the lecture begins, these worries evaporate. Professor Ruttner's lecture is committed, competent, sober and scientifically demanding. Just like Niko loves. No doubt, here is a scientist who is fully involved in the subject and reports on the latest results of research. Niko is reassured. Such a man can also do future-oriented science in a shack. Gudrun is also taken with Professor Ruttner. She particularly likes his Austrian charm, the Viennese tips of words with which he peppers his lecture, whenever something is particularly important to him. Gudrun is happy for Niko. Has he found a good mentor here?

After the lecture, Niko goes to the front: "I have been interested in honey bees for many years and have experience in beekeeping. Can you use a student assistant?"

Professor Ruttner is pleased and lapses into Viennese: "Well, look here. Oan Jungimka! Gladly! Yes, welcome, I can employ you on an hourly base as an assistant."

He laboriously rummages his calendar out of his large briefcase.

"Let me look."

Then again professor in High German: "Next week Tuesday looks good. Nine o'clock? Then we can discuss the details in Oberursel. I can also introduce you to our master beekeeper Mr. Blotz. He will be happy about help."

Just in time for the agreed appointment, Niko rings the doorbell of the Institute in Oberursel. A young woman opens the door and asks in a cheerful Austrian accent what he wants.

"Yes, the professor is here. He has a phone call right now. Can you wait a moment?" and disappears again.

Niko stands awkwardly in the small, winding hallway until Professor Ruttner calls him in: "It's nice that you found your way here. Tell me: How did you learn about bees?"

"From my father's beekeeping!"

This is the start of a longer conversation about the merits of bee research in general and the professor's specialty, beekeeping and mating control in particular. In the middle of the conversation, the young woman who has received Niko comes and brings coffee. Professor Ruttner introduces her to Niko:

"This is Miss Agnes Bachler, who came with me from Lunz as a technical assistant!"

After coffee, they go to the garden behind the house, through the apiary and back through the cellar entrance. In the main basement, a narrow wood workshop has been set up in one corner. Master beekeeper Mr. Walter Blotz is working on a board.

"No circular saw?" asks Niko cheekily.

"What's the point of a circular saw when you can handle a foxtail saw well?"

Master beekeeper Mr. Blotz has obviously not fallen for his mouth, takes his board and screws it to another on the small workbench. He has grown tall and reaches almost to the basement ceiling.

"So, you have experiences in practical beekeeping? Where did you learn?" he asks doubtfully.

"I helped my father with his eight bee colonies. Now I'm looking forward to learning more about working with bees."

"Hmm, eight colonies, after all."

When Mr. Blotz hears that Niko can only help one morning and one afternoon a week because of his studies, he frowns: "So irregular?"

"What do you have planned for Thursday morning?" asks Professor Ruttner. "And, Niko, can you be in Oberursel in the morning?"

In the morning is 7:30. It is still dark when he has to get up. And although Niko arrives at the Institute on time, Mr. Blotz is busy loading the Institute's VW bus. Niko gets a beekeeper's hat with a veil and off they go. The first apiary is very close, an apiary with more than 30 colonies. Mr. Blotz explains: "Today we are working on swarm prevention! A close inspection of each colony and give more space so that they lose their swarm instinct. That's why we give empty combs or foundations to all colonies! And everything must be noted on the "hive card" of each colony!"

Niko's offer to control colonies as well is harshly rejected: "You can't. The Professor said you should give me a hand!"

Niko's job is to operate the bee-smoker so that Mr. Blotz has both hands free. As soon as the first bees start stinging flights, Niko gives the first puff of smoke.

"Not yet!" comes impatiently from Blotz. He waits for his command from the next colony. The reaction: "You have to smoke earlier!"

Niko knows this from his father—he has the choice "between too much and too early" or "too late and too little"! Years later, Niko realizes that the time when smoke has to be given is inevitably always wrong. If the bees have not yet stung and then the smoke comes, it is "too early"! If you get stung and the smoke comes afterwards, it's "too late"!

Nevertheless, Niko learns a lot, just by watching. There is a difference between having to work on only eight colonies and over a hundred. The beat rate and efficiency are completely different from what Niko knows from his father. Nevertheless, he does not warm up to Mr. Blotz. He seems to have a general problem with students, who are all good for nothing and are chasing some fantasy. Apparently, Mr. Blotz passes on his displeasure about the additional workload of supervising a student to Professor Ruttner who circumnavigates the problem diplomatically and finally summons Niko at the weekend to look after the test colonies together with him.

The year 1965 is a good honey year and soon boxes filled with honeycombs are piled up in the apiary. Blotz explains to Professor Ruttner that he can't do it alone and that he doesn't have time to extract honey anyway if he is supposed to look after all the test colonies.

"Niko, we're drowning in honey this year. Could you help to extract and work a few more overtime hours over the next few weekends?"

Niko is delighted. He senses the chance to introduce Gudrun to the Bee Institute.

"Yes, I can extract honey during the weekends. It can hardly be done alone. Could I bring my friend Gudrun to support me? She's also a biology student."

"Of course, you are welcome to bring Miss Gudrun with you. Does she like to work overtime on weekends?"

Professor Ruttner is happy to be able to get another student's help so easily. On the one hand, students are inexpensive, highly motivated, and on the other hand, he can bind young scientists to his research group in this way. Gudrun is thrilled to be able to help with the honey extraction. It reminds her of her grandfather, who died at an early age. So, Gudrun and Niko work together in the beehouse in Oberursel and try to get the towers of boxes of honeycombs under control. Again, there is an effective division of labor between them. Gudrun quickly learns to open the honey cells, which have been sealed with wax by the bees, with a special fork, the "uncapping fork." A terribly sticky affair, and despite the long white lab coats, the clothes are not spared. The hands stick, pants stick, the shoes stick, everything sticks. It's fun to see how the golden-yellow honey runs from the centrifuge through the honey sieve into the bucket. Niko loads the extractor, changes the buckets and cleans the sieve. Things are progressing well. Professor Ruttner, who comes to them in the apiary from time to time, is visibly satisfied. "Well, it's like piecework here."

He is then interested in Gudrun's work and he engages her in conversations about her studies and biochemistry. This long conversation distracts Niko's and Gudrun's attention. As a result the honey bucket overflows, and the sticky mass spreads on the wooden floor of the beehouse. "Oh dear. What have you done now?" comes from the Professor. "You know, honey is very sticky. Nothing helps now but wiping up with a lot of water." He smiles pityingly and shows the two of them where the floorcloth, mop and cleaning bucket are. Sticky, they had noticed that before. They are now learning how difficult it is to remove the honey completely from the wooden floor. Simply impossible. Not even with hot water, not even with abrasive cleaners. It remains sticky. Only after a few weeks is the beehouse accessible again without a crackling sound of shoes. Niko does not ask Professor Ruttner how master beekeeper Blotz classifies this episode.

The next day, Professor Ruttner is very excited. "Niko, in two days Professor Karl von Frisch will visit our Institute. He will see Professor Lindauer in Frankfurt and then comes to us. Please also let Gudrun know that she should be at the Institute too." Niko interrupts him: "The famous Karl von Frisch, who deciphered the bee dances? Of course, Gudrun and I would love to come. Thank you very much for the invitation. We really look forward to meeting this most illustrious eminence of honey bee research."

Fig. 3. Professor Karl von Frisch with his daugher at Institute of Bee Research in Oberursel, 1965.

In the early afternoon, Gudrun and Niko, together with Horst Herrmann, who has just been accepted to start a doctoral thesis, are eagerly waiting for Professor von Frisch. Finally, they see a car stop in front of the Institute. Professor Ruttner hurries to greet Professor von Frisch and his daughter very warmly and disappears with them into his study. It doesn't take long and all three come into the breakfast room and Professor Ruttner introduces them as his students, who probably all want to do their doctorate with him. After a few questions about why they were interested in bees, von Frisch congratulates his host for being able to arouse the students' interest in bees in such a short time. Afterwards there is a guided tour of the Institute and the bee garden. Above all Niko is deeply impressed.

After that, Gudrun, who has succumbed to Professor Ruttner's Viennese charm, comes to the Bee Institute more often and helps wherever necessary. In June, Professor Ruttner asks if Gudrun and Niko would help him with experiments in Austria for two weeks after the end of the semester. "Unfortunately, I don't have the money to pay you by the hour like here in Oberursel, but room and board would be free." Niko is immediately enthusiastic. Finally, there is an opportunity to get to know bee research directly. Gudrun wants to know more about the experiments and what role they should play in them. Professor Ruttner patiently explains: "This is part of a large project in which we want to understand the mating behavior of honey bees. You know that the queen mates with many drones on a nuptial flight at high altitude." Gudrun had heard that in his lecture. "That's why no one can control the mating. This is a big problem for beekeeping, where you naturally want to mate breeder queens with specific drones. Just like elsewhere in animal breeding. You may have read that I have been working on this topic for many years and use the Limnological Station in the municipality of Lunz am See as a location."

No, Gudrun didn't read that. Professor Ruttner ignores it and continues enthusiastically:

"There we know drone congregation areas to which the queens fly to mate. This year, we want learn the specific colony origin of the drones in the congregation area. To do this, the drones in all apiaries in the valley must be marked with different colors. We then catch the drones again at the drone congregation area, note their colors and thus know from which apiary they come."

Professor Ruttner beams. He is only too happy to talk about his experiments. Even if it sometimes takes a little longer than the listeners would like. Gudrun understands the experimental concept and is convinced. "Okay. That sounds exciting. I'm in, my internship in biochemistry is only in the second half of the semester break."

She looks at Niko, who nods happily. He would have agreed. He was worried that it would be too little biochemistry and too much field zoology for Gudrun. Professor Ruttner explains the details of the best route to get to Lunz and where exactly they live in the Limnological Station: "Excellent! Then we will meet this summer in "Lunz am See". I look forward to our cooperation."

Gudrun and Niko have never been to the Alps before and suspect that their Freiburg equipment might not be enough. On Professor Ruttner's advice, hiking boots suitable for mountaineering and new waterproof windbreakers are purchased. The main problem: The Heinkel scooter is getting on in years. He had a hard time in the mountains of the Black Forest. And now to the Alps? Niko no longer trusts the old vehicle. He does not want to be left stranded on an Alpine pass with a damaged cylinder head. And Gudrun still remembers well how she offered Niko to help push so that Heinkel could manage the climbs.

Their financial resources are limited. It's not enough for a car. The West German "economic miracle" comes to their aid. Many people are now in a better financial position than they were a few years ago. A car boom sets; motorcycles and scooters are now offered like sour beer. They rummage through the advertising section of the Frankfurter Rundschau. The search is arduous. They drive the Heinkel all over the city in search of a suitable replacement. Nothing they look at fits: too expensive, too used or both. In the end, Gudrun discovers the best offer in a workshop around the corner: a Horex Regina Sport 350cc, a strong and large motorcycle. 20 hp! The chrome-plated gas tank shines in the sun, the aluminum cylinder head shimmers matte. The price of 320 Deutschmarks seems surprisingly cheap. The vehicle even has a comfortable sprung pillion seat and two large panniers. Ideal for long trips. Gudrun is thrilled. The next day, Niko also takes a look at the motorcycle: "Could I take a test ride?"

"Of course, you don't want to buy a pig in a poke. Why don't you come by tomorrow at three o'clock. Bring your driver's license and crash helmet!"

They are ready punctually the next day and Niko tries to start the machine. However, the kickstarter does respond. The master looks pitying and Gudrun smiles mockingly.

Then the foreman points to the lever on the steering wheel to reduce compression. He winks at Gudrun: "Well, my dear, a motorcycle is like the girls. You can't be too shy. So vigorously. Something like that." The master briefly steps on the kickstarter with all his weight, and the machine immediately runs like clockwork. Satisfied, he then switches off the engine again. Now it's Niko's turn. Just don't embarrass yourself. He concentrates and presses his right foot on the lever with all his might. The Horex wakes up with a sonorous engine sound. Niko beams and swings into the saddle. He accelerates and enjoys the vibrations of the powerful engine. First gear and off with a lightning start. Oops, the clutch. Gudrun is glad not to have to go along right away. Niko first has to get used to this acceleration. He drives over the "Alleenring". After every red light, he is first! Then the Opel roundabout. With perfect cornering on the motorway towards Kassel. Change to the left lane, which immediately becomes a fast lane for Niko on the Horex. Only now, suddenly, actually too late, does his rationality regain the upper hand! Over 130 kilometers per hour on a strange, completely unfamiliar two-wheeler? That is not possible at all! He gets out again at the Bad Homburg exit and returns to Bockenheim at a VW Beetle pace, well below 100, where Gudrun is worried and the foreman is angrily waiting. Niko tries hard to hide his enthusiasm. He asks cautiously:

"The price is affordable. Does the machine have any defects or quirks?"

"Oh no. The machine has to go. My friend Mischi died two weeks ago in the thunderstorm, the wind pressed him against a bridge pillar. The Horex came through without serious damage. Just a new front wheel, the fork and the headlight. If only it had been other way around." The man is visibly working on his composure and wipes away a tear. "Take it with you and just be careful with it. Nobody wants more casualties." Gudrun and Niko have a queasy feeling. In any case, a drastic reminder to exercise caution when riding!

On Saturday, they will start with Horex to Austria. Packing is much easier than with the Heinkel, a lot fits into the side pockets. The small tent is fixed under the headlight, the rest ends up in the backpack on Gudrun's back, as usual. They start early in the morning. It's too cold for July. The newspapers that they have shoved under their jackets to protect themselves from the wind do not keep them warm enough. Shortly before Nuremberg, the first break is taken for a warm-up gymnastics.

As soon as they are back on the highway, it starts to rain. As experienced motorcyclists, Niko and Gudrun stop under the next bridge. Time passes, and the rain becomes even heavier. Niko gets nervous, they absolutely have to get to Lunz today.

"Shall we continue?" "I come from the North Sea coast. I don't mind rain. I'm wet anyway and I'm certainly not made of sugar." The rain is getting heavier and heavier and the visibility through the wet motorcycle goggles is not improving. They take much longer than planned. Behind Passau, they cross the border into Austria. Still in the rain. The customs officers do not control anything. They just laugh pityingly and ask: "Where do you "two water mice" want to go?" "To Lunz." "Well, it's not far to Linz."

They are "waved through". Niko nods friendly. He knows he has to pass Linz to reach Lunz. That's another more than a hundred kilometers through the rain. Uphill and downhill over winding country roads. Fun is different. At least it's not very cold. Completely soaked, they finally reach the longed-for town sign "Lunz am See". It is dark when they refuel the Horex at a gas station at the entrance to the village. The gas station attendant looks at them sympathetically and offers them to come in to dry before continuing their journey to warm up a little. "Oh no. Thank you very much. We only have to get to the Limnological Station! That's not far away now, is it? Do you know how we get there?" "Yes, that's not far. Just follow the road to the right. Along the lakeshore and then up the slope."

In the dark and in heavy rain, only walking pace is possible. After what feels like an endless drive on a narrow, winding road, they finally reach the station. There, the lights above the front door shine brightly. An elderly couple, who introduce themselves as "Aunt Pepperl" and "Uncle Fritz" of "Miss Agnes" from Oberursel, warmly welcome Gudrun and Niko. The people from the gas station had called them and announced a wet German pair on a motorcycle. In Lunz am See, they soon realize, everyone knows everyone. Aunt Pepperl was a little worried because it took so long. Uncle Fritz pretended to have predicted everything exactly, but was glad that the students from Germany had made it to Lunz in one piece.

First of all, Uncle Fritz gives Gudrun and Niko a high alcohol fruit brandy: "That helps. Otherwise, you'll catch a cold!"

A bathtub with hot water is prepared, so the only remaining dry part of the body, the scalp under the motorcycle helmet, also gets wet.

Liquor and hot bathtub did the job. At dinner, they can only keep their eyes open with difficulty. Then it's off to bed. They are so tired that they only curse softly about the separate accommodation, Gudrun with the women and Niko in the men's bedroom.

Third chapter

Gudrun and Niko catch drones in Lunz and help build a house on Lake Neusiedl

The next morning, Niko wakes up late after a long deep sleep. Gudrun was awake earlier and conjured up a breakfast for them. They start the day slowly. The sun shines through the window. The alpine summer retreat is like a photo in a picture book! It is no vacation, Gudrun and Niko are supposed to work here.

At noon they have an appointment with Professor Ruttner in the nearby restaurant "Seehof". He buys them a good lunch and explains to them once again full of enthusiasm the importance of the upcoming experiments.

"Schauns (Look), I have bred bees here in Lunz for many years, which produce a lot of honey and sting little. I would now like to breed them in Oberursel and also pass them on to many beekeepers in Germany. They will certainly be thrilled when they get to know them for the first time. For breeding, I have to ensure controlled matings of the queen. No problem with vertebrates, you simply lock the pair together in a stable. Not the honey bees! The queens fly out for their nuptial flight, they simply fly high and are no longer seen. They don't come back until 20 minutes later. Then they mate with several drones. They have mated. Who did they mate with? In order to be able to control these matings, we need to know much more about the behavior: Where do the queens fly to and which drones do they hit? How far do they fly? We want to find out all this now!"

Niko has read up on drone flights and is skeptical: "We know that the daily flight period of the drones starts around 2:30 p.m. and ends at 4:30 p.m., when the temperatures remain above 18°C and the sun is shining." "Yes, but we don't know how the drones are distributed and how far they fly. Let's explore that now! To do this, we will dab a patch of paint on the drones in all apiaries with a brush. Each apiary gets a different color! You must carefully log the number of marked drones. Particularly important! Around

10 a.m. we have to stop marking! The drones become restless at this time and then fly away from us. When we have made it through all the apiaries, we want to catch as many marked drones as possible and through the color we know where they come from, how far they have flown and how they are distributed. From previous experiments, we know where the drones gather in search of queens ready to mate. That's why these places are also called drone congregation areas! First of all, we have to check whether these places still exist."

At the Limnological Station, there is another briefing for the entire motley group of helpers. Professor Ruttner's brother Hans, the "Mr. Engineer", is involved in the experiments. He heads the bee department after his brother's move to Frankfurt. In addition to two beekeepers from the station, Hans Ruttner's three children and his student Hermann help out. In addition to Gudrun and Niko, Professor Ruttner's wife, Dr. Sophie, and Miss Agnes Bachler have come along from Oberursel.

"You still know from last year what it's all about. We mark the drones in the early morning, then at 1 p.m. we all meet in the courtyard of the station to determine details of the operation and the respective assignment of the people. Then the balloons will be filled with hydrogen and finally it's off to the drone congregation area to see if the drones will fly there again this year." Getting up early is difficult for Gudrun and Niko. After a few days, they are "wide awake" around seven in the morning and, with their paint brush in their hands, hunt down drones that scurry around among the bees on the comb. One apiary after the other is done. White, blue, green, red and finally yellow drones can now be assigned to their respective apiary. In the afternoon, they check whether the two drone congregation areas from last year can be confirmed again this year.

At 2 p.m. sharp, Gudrun and Niko are at the drone congregation area "Seekopfsattel". Professor Ruttner instructs them to keep very quiet so that they can hear the sound of the drones flying. They wait and nothing happens. Gudrun is thinking of giving up when suddenly, after half an hour, a buzzing sound is heard. First softly, then louder and louder. The Professor's trained eye can see the first drones flying. Gudrun doesn't see anything, but the sound is now an overwhelming "roar". They let a balloon rise on a string to about 20 meters. Below the balloon, a queen bee is attached in a cage as bait. Her typical scents attract the drones, which now form dense drone swarms like comet tails. They dissolve just as quickly as they formed, and then form again and again. Gudrun and Niko are impressed by the spectacle. "Gudrun, look! That's at least 40, sometimes maybe even 100 drones." "Oh, really? I had actually only counted a maximum of 87."

"They fly really high, certainly 20 meters. How are we supposed to catch them?"

This year, too, they are attracting hundreds of drones at the two drone congregation areas "Lochbachwiese" and "Seekopfsattel", while no drones are attracted at other checkpoints.

After four days of drone painting, the experiments can begin and the marked drones can be caught. This operation looks bizarre to bystanders. The "pilot" tries to steer the balloon to the appropriate height. The drone catcher has a four-meter-long pole with an insect net attached to the tip (Fig. 4). The balloon is then slowly pulled down to lure the drones down to the height of the insect net so that they can be caught. Gudrun is the pilot, Niko the catcher. Once a large swarm of drones has gathered at the queen's cage, Gudrun carefully pulls the balloon with the swarm of drones down as evenly as possible to within reach of the net. Catcher Niko rushes over and tries to catch as many drones as possible with a well-aimed blow of his net. This procedure proves to be difficult. Often, the drones do not follow the queen on the balloon down to the catch height. Gudrun has to let the balloon rise again and wait until enough drones have gathered once more. Then the game begins again. Success is by no means certain. However, the height of the balloon depends not only on the length of the balloon string, but also on the wind speed. When the wind dies down, the balloon rises higher with the queen and drones. If the wind picks up again, the balloon is pushed down at an angle. So, the catcher has to run back and forth and many well-aimed blows with the net goes past the drones, which many bystanders (including Gudrun!) find amusing. Niko gets the nickname "Don Quixote" because he keeps fighting against nothingness with his long lance.

Fig. 4. Catching drones at the Seekopfsattel attracted by the caged queen high in the air.

Professor Ruttner, his wife and Agnes have also come to the Seekopfsattel to get the drones out of the net and record the colors. To their delight, many of the captured drones are marked. All drones, marked and unmarked, that are caught at the Seekopfsattel now get a yellow mark on their abdomen (Fig.5). At the Lochbachwiese drone congregation area (DCA), they are marked in blue. For example, the flight distance and distribution of the drones are determined for the various apiaries on the Seekopfsattel day by day.

The result is a uniform picture. The drones mainly come from apiaries at a distance of two to three kilometers. Drones from the two apiaries four kilometers away are rare. In most cases, drones are increasingly represented by nearby colonies. After two weeks, all colonies on all apiaries are checked again. All drones that have been previously marked on one of the two drone collection areas are collected from the colonies, sorted according to their splashes of color, and the respective number per colony is carefully logged.

Helpers are invited to the final meeting in the Seehof restaurant. After enjoying fresh trout, delicious dumplings and plenty of wine, Professor Ruttner reports on the results

Fig. 5. Professor Ruttner and his wife mark drones caught at the Seekopfsattel. All drones, marked and unmarked, get a yellow mark on their abdomen.

in the best of moods: "You did very well. Thanks to your cooperation and the good weather, we were able to collect a lot of data. We will evaluate them in detail later. I can say in advance: The vast majority of drones remained faithful to a particular drone congregation area. Only a small proportion of the drones switched between the two congregation areas. It seems as if most of them only fly two kilometers, but some flew up to 4 kilometers. For beekeepers, this means that they have to control all drone colonies in a diameter of about 8 kilometers to be sure of mating control."

Niko sees this as a big problem for the beekeepers. He quickly crosses the circle area and speaks up: "Professor Ruttner, that's 50 square kilometers. A beekeeper can't control that, can he?"

"Yes, that's right. A real breeding problem. We will therefore collect data again next year to show that our values from this year are repeatable and therefore scientifically robust. Then we will discuss this with beekeepers. So thank you again—and cheers to next year. I very much hope that you can all be here again."

Now that the experiments at the drone collection areas for this year have been completed, Gudrun and Niko treat themselves to something as exotic as a vacation for a whole week for the first time! Without bees, without chemicals, just Gudrun and Niko. The Horex shares the good mood and purrs leisurely through the Austrian mountains. Their destination is Illmitz on Lake Neusiedl. Gudrun insisted on "holidays without bees" and asked engineer Hans, Ruttner's brother, about the best place without bees and without beekeepers. He should know. He had to think about whether such a thing even existed, then he suggested Illmitz.

Looking for a place for her small tent, Gudrun asks at a farm in the village. Yes, they are allowed to camp on a meadow by the lake. The old farmer describes the way to Gudrun and before it gets dark, Gudrun and Niko lie comfortably in their sleeping bags. The Horex is well lashed and covered next to it. The plan is to get up late, only when hunger wakes them up doesn't work At half past seven they hear loud voices in front of the tent and then a rattling, roaring machine. Gudrun growls wearly: "Niko, take a look, see what is going on. I still want to sleep!"

Niko laboriously peels himself out of his sleeping bag and crawls to the tent entrance. Carefully, he opens the zipper of the tent entrance and sees a concrete mixer rotating vigorously a few meters in front of their tent. People with shovels and hoes are preparing a construction site. The old farmer had neglected to warn them."Gudrun, we have two options. We can get angry, pack our things and drive away." "And the second?" "I haven't worked in construction for a long time. What do you think?" "I have to sleep! Leave me here in my sleeping bag and do what you want. I can't help you anyway!"

After all the minutiae marking bees, Niko is looking forward to some hands-on work: "Good morning, it looks like you could use a helper. I can take over the wheelbarrow and bring you the gravel from the pile!" People look at each other in surprise. They can use a helper. Around 9 o'clock a horse-drawn carriage arrives with breakfast. Now Gudrun is also there. They set up two benches and a table. There is bread and bacon. The cups have been counted and are not enough. No problem: Gudrun fetches the large coffee mugs. There is no coffee, but yogurt and two large bottles of wine. The man for whose family the house is being built is the driver of the dairy vehicle. He brings the milk from Illmitz to Vienna and from there the finished products back to Illmitz. As everyone laughingly assures them, they never drink coffee here, but only wine. The aromatic and slightly sweet "Gewürztraminer" grow in the fields around and is pressed by the farmers themselves. For Gudrun and Niko, wine for breakfast is unusual. The sun is burning from the sky and they are thirsty. After the first two glasses, they are convinced that this wine in the morning will have a very good effect. Gudrun laughs and promises: "In the future, only wine from Illmitz will be included for breakfast in Frankfurt Bockenheim!"

The people are happy and promise: "As long as you are here, there will always be enough wine." Then the work continues. Gudrun stands with a shovel at the mixing machine. Three shovels of gravel and a sack of cement! Niko strugggles with a heavy wheelbarrow filled with gravel! Around twelve o'clock there is a break. There is a spicy soup with lots of meat, called "Geselchtes", and bread. And again, the good wine. Niko is amazed to see that Gudrun dilutes her wine with water. So far, he has always considered Gudrun as a "hard-drinking" woman who never refuses a good glass of wine. They sweat in the blazing sun and Niko's head is also painfully noticeable. So, Niko, despite the ridicule of the locals, also reaches for the water bottle. Nevertheless they both make it to the end of work day. They politely decline an invitation to dinner in Illmitz. They need a good night's sleep so they crawl into their tent. It is dark after a few hours, and they wake up and are hungry. In front of the tent, they find a pallet of yoghurt, two bottles of wine and a large piece of bacon. Thus, they enjoy a rich dinner, once again with wine! The efforts of the day are noticeable. The next morning is Sunday and they can sleep in late. Only when the sun shines strongly on the tent does Gudrun quietly crawl in front of the tent. This time there is no wine for breakfast, but mugs of hot coffee. She starts the gas stove, and when the water boils, she puts two spoonsful of Nescafé in each cup. The scent wakes Niko, he laughs: "No wine? Gudrun, what has become of you?"

The bird life at Lake Neusiedl proves to be unimaginably rich! A large shallow lake with huge reed beds, swampy soils full of insects and mosquitoes offers breeding birds, migrating birds good protection and sufficient food. With binoculars and a little patience, they discover many bird species, including rare ones, over the next few days. Satisfied, they drive back to Frankfurt.

Fourth chapter

Gudrun and Niko start their doctoral theses and marry

"Biochemistry is about the really important mechanisms that control life. We can carry out controlled experiments without disturbing winds that drive the drones away." Gudrun doesn't even wait for Niko's answer. "The experiments in Lunz were fun. Lots of exercise in the fresh air. What about in winter? Niko, what does the bee researcher do in winter? The bees hibernate, so there is nothing to do. That's another big handicap in bee research!" Niko avoids a clear answer. He talks about preparations for the coming season, but can't convince Gudrun. She is registering for the biochemical internship for the winter semester.

A few days later, Niko comes with a new message. He was at the Bee Institute and Professor Ruttner asked if they would be willing to do disease diagnosis on bees. The Hessian Beekeepers' Association wants to know how many bee colonies were infested by Nosema, an intestinal disease in the adult honey bee. The beekeepers collected dead bees from colonies and send the samples to the Institute. These bees must be examined microscopically by March. The Association pays well for it: 50 pfennigs per sample. Gudrun has decided. Niko will have to examine bee samples without her. She did not keep to her decision and on the weekends, Gudrun often worked together with Niko at the Bee Institute. The examinations proceed quickly and the earnings are impressive at the end of the day. Nevertheless, the pile of samples sent in is not getting any smaller. Professor Ruttner now also assigns Agnes and a little later also his new doctoral student Horst Berrmann.

Niko is rummaging through a new stack of samples when the Professor enters the lab. "Can you interrupt your work for a minute, Niko? I have a concern that I would like to discuss with you in my office." Niko is surprised and also a little worried. In the office, Professor Ruttner sits down behind his desk: "Dear Niko, I have had you here at the Institute for almost a year and I am very impressed by your interest and enthusiasm for bees. That's why I would like to suggest that you start a doctoral thesis here next spring. I have a topic for you that interests me very much. A question that has been debated for more than 150 years concerns the queen bee's ability to distinguish between the small worker comb cells in which she lays worker eggs and the larger comb cells for drone eggs. You will need to find out how the queen recognizes this!"

"That sounds interesting! I will read the literature on this and let you know if and how I can tackle the problem." "Of course, I can always help you!"

The Professor hesitates for a moment before continuing: "I would also be glad if Miss Gudrun would start her doctoral thesis here. I know that she is mainly interested in biochemistry. That's why I also prepared a biochemical topic for her. Please let her know that she should come to see me one day!"

In the evening, quite late, Niko stands in front of Gudrun's door. As far as Gudrun can understand in the confused excitement, Professor Ruttner has asked if he, Niko, wants to do a doctoral thesis with him at the Bee Institute. And also, he asked Gudrun to come to Oberursel next Saturday. "Niko, you're crazy! We are now in the 6th semester. Our studies are far from over, we are still missing several internships. We won't be able to talk about a doctoral thesis for two years at the earliest!" Niko remains unimpressed by Gudrun's objections. He is determined to do his doctorate under Professor Ruttner: "Six semesters or not, Gudrun, this is our great chance to get started in bee research. We have to take advantage of this opportunity!"

Gudrun's landlady shouts through the wall: "It's late! I ask for silence!"

"Gentlemen's visit" after 10 p.m. is prohibited according to the rental agreement. Gudrun's landlady had said before she "is not like that", but there is a paragraph that talks about "matchmaking", and she is afraid of that. The two quickly say goodbye and arrange to meet the next afternoon, after Gudrun's internship with Niko in Kleine Seestraße.

When Gudrun arrives, Niko talks a little more calmly about the topic that Professor Ruttner suggested to him. How does the queen know which eggs belong where? Gudrun is shocked: "The chances of success for this project seem very low to me. And who cares about this question?" "Let's first hear what Professor Ruttner proposes in detail!" "Niko! I'm worried that you'll get lost in your enthusiasm for bee research and end up failing. I don't think it's wise. I don't think I should let myself be dragged into it." That can't deter Niko. And he manages to get Gudrun to at least keep the appointment with Professor Ruttner: "A discussion can't hurt".

On Saturday afternoon, she takes the tram to Oberursel. The welcome is warm: "Has Niko told her about his doctoral thesis?" "Yes, he's all enthused"

"Miss Gudrun, you know, I would be very happy if you would also come to the Bee Institute as a doctoral student. Bees cannot do without biochemical processes either. You know that." Gudrun knows. Professor Ruttner clears his throat: "And there is a question that has been burning under my nails for years, but which can only be solved with competent biochemical expertise!"

Professor Ruttner continues: "It's about the years of storing the drone sperm in the queen. The queen only makes nuptial flights at a very young age, after which she can fertilize her eggs with the sperm for up to five years! A topic that we have talked about several times in Lunz! How does the queen manage to keep the sperm alive for years in a special bubble, the spermatheca? Does the queen have to nourish them or does she simply keep them inactive as if in some kind of sleep? Does the large gland on the spermatheca have anything to do with it?"

Gudrun is all ears. That sounds very interesting indeed. How do sperm survive for years in queens, but only for a very short time in the females of most animals? Gudrun, however, does not forget her basic reservations:

"Niko and I have only studied for six semesters. We still have to do a number of internships before we can start with a doctoral thesis. It's just too early for Niko and me!"
"Have you studied the doctoral regulations at this university? That always helps with such formal concerns. At no point did I read anything about a minimum number of semesters as a prerequisite. Have you?" "I didn't even know that there was such a thing as doctoral regulations." "Look. Sometimes it is worthwhile to study not only biology but also the regulations. Just give me your study books and documents. I then want to clarify with the faculty which lectures and internships still need to be completed before the doctorate. Besides, it can't hurt to start your doctorate as early as possible? Keep

in mind that such a doctorate can sometimes take unexpectedly longer. I know you well enough to know that it won't be long before you can show nice results!"

Professor Ruttner radiates such convincing optimism that Gudrun does not dare to raise any further objections. She is not quite comfortable with all this. She feels taken by surprise. And Gudrun would like to read the doctoral regulations herself. She asks for a week to think about it and at the same time for literature, so that she can find out what has been done on the subject. Professor Ruttner is visibly satisfied with the outcome of the conversation. He shuffles to a large shelf in his office. There he has archived copies of published research papers, so-called "offprints". Awkwardly, he fingers out several works and hands them to Gudrun:

"However, I would like to have them back. The special print collection is a very important basis for academic work. You don't always want to go to the library in Bockenheim. For your dissertation, the first step is a literature review on the topic and then a written work plan. We should have the work plan ready before the start of the bee season so that we know how many bee colonies we need for it!"

On the way back in the tram, Gudrun reviews the conversation. She is irritated by Professor Ruttner's "we." As far as she knows from friends, literature development and work plan are achievements of the doctoral student that decide on the official admission to the doctorate.

Niko is certainly eagerly awaiting her report, but that is not possible for the time being. Has he informed himself about the doctoral regulations? It's all too much for Gudrun at once, she has to find herself again now. So, she doesn't get off at the Westbahnhof in Bockenheim, but continues by tram to Frankfurt Zoo, for which Niko and Gudrun have an annual ticket. The green oasis in the city centre is an important place for them to sort out ideas and find peace.

The discussions with Niko later go as she expected. His enthusiasm doesn't help, but this argument: "We won't lose anything if we start doing our doctorate in Oberursel now and at the same time do the internships and certificates that are still necessary. We notice whether the doctoral thesis is going well or badly after two years at the latest, but probably much earlier. Even then, we will still be well on time."

She gives herself a jolt and Niko a kiss: "Let's go, Ph.D. student!"

Gudrun's literary work is more difficult than initially hoped. The offprints are all a bit older and do not deal with biochemistry. Professor Ruttner cannot answer her numerous questions about chemical analyses. She talks to Professor Pfleiderer about the topic during her chemistry internship. He finds the question interesting, but has concerns:

"How much fluid is there in such a spermatheca?" "About a microliter."

Pfleiderer looks up in shock: "Just a microliter? Can they get enough of the spermathecas to perform such analyses? Think of how much liver you have to put through the meat grinder to get a few milligrams of the enzyme lactate dehydrognase."

He reflects: "Even if we assume a high concentration of the ingredients, this is far too low an amount for all the analytical methods that I am aware of! They will need a lot of queens!" Then hesitantly: "On the other hand, there is now a new method for separating small quantities, called thin-layer chromatography. The separating agent is not poured into thick columns as before, but thinly spread on glass plates. Maybe, it could work that you separate the individual ingredients of the spermatheca. As luck would have it, I was just able to buy such a machine. You could practice with it. We don't have any experience with that yet. If you succeed, we could test in the spring how many queens you need so that we can see something. In principle, I am happy to be on board with the topic."

Gudrun thanks Professor Pfleiderer for the suggestion. Now, in addition to the lectures and the literature work, she produces thin-layer plates for the entire working group. Soon she manages to apply the layer evenly. Pfleiderer is enthusiastic. Now she would have to produce queens on a large scale. Professor Ruttner generously assures her of his full support when asked. "Of course, you will get enough queens for your examinations!"

Niko also buries himself in books, research literature and even in old bee magazines. At their daily meetings, Niko only tells her who wrote what and about when the distinction was made between the two comb cell types. He also presents to her the opinion of the famous "bee pope" Johann Dzierzon, (Fig.7), which is absurd from Gudrun's point of view:

"The queen has "free will" and decides when laying eggs that a fertilized egg belongs in the worker cell and an unfertilized egg in the drone cell!" "Funny, what is "free will" supposed to mean? It's just a matter of how the queen can distinguish between the two cell types."

Niko reacts angrily to this statement: "With the state of knowledge around 1860, "free will" means that, in Dzierzon's opinion, the queen is able to control whether an egg is fertilized or laid unfertilized." "Yes, the sperm somehow live for years in the queen's spermatheca, the organ I'm supposed to work on. And yes, there is a complicated valve at the exit of the spermatheca, which is innervated by the central nervous system of the queen."

Maybe old Dzierzon is right after all, Gudrun thinks and asks: "What did this Dzierzon say about how the queen distinguishes between worker and drone cells?" "There is nothing to be found in Dzierzon's publications. That's my topic. That's exactly what I'm supposed to find out!"

Gudrun longs for a break and a short relaxation: "What do you think about a trip to the zoo? Without a word about our literature work, the doctoral topics and bees. Agreed?"

Fig. 7. Dzierzon who stated that the queen has "free will" and decides when laying eggs that a fertilized egg belongs in the worker cell and an unfertilized egg in the drone cell!"

Niko enthusiastically agrees. For him, too, the zoo is a place of relaxation and inspiration. They arrange to meet the day after tomorrow, next Wednesday.

Unfortunately, Wednesday is rainy and cold. Not a good day for a visit to outdoor enclosures. Off to the zoo's Exotarium with the aquariums and terrariums! Niko knows most of the fish, because aquariums have fascinated him since his childhood. He can't resist listing the names. The richly equipped underwater worlds challenge them to extensive search games to find the animals on the display boards. Then they go up the stairs to the reptiles and insects. Niko heads specifically for the rear area. Isn't there an observation hive with a honey bee colony. What was the agreement? Should Gudrun protest? That would help little and only ruin their day together. Gudrun walks slowly on, marveling at the colorful poison dart frogs and the large colony of leafcutter ants. Niko has his nose on the glass pane of the observation hive. He seems to have forgotten everything around him. Gudrun wants to wait until he reappears from his "honey bee universe." Suddenly Niko comes to her excitedly: "Please Gudrun, look at the queen who is laying eggs. Before laying eggs, the queen inspects the cell. She bends down into the cell up to her chest ring. And please, look closely! I have seen that the queen inserts both front legs into the cell (Fig.8). This is completely new! This is not in any of the many descriptions of the cell inspection. No one recognized this and described that it puts its front legs into the cell in addition to the antennae. Please look, here is the queen!"

Gudrun, although slightly annoyed, sits down in front of the observation hive and watches the queen. She can clearly see that both front legs are also inserted deep into the comb cell. Then the queen pulls her head and legs out again, walks a little ahead and lowers her abdomen into the previously inspected cell, probably to lay the egg. Niko is happy about Gudrun's confirmation. Intoxicated by his "great discovery", he

Fig. 8. Before laying eggs, the queen inspects each cell. She bends down into the cell up to her chest ring at the same time inserting both front legs and both antennae.

begins to philosophize: "How many bee researchers before me have observed a queen bee laying eggs inserting her front legs into the cell as well as her antennae. No one noticed what we both just saw, dear Gudrun!"

Gudrun should have known better. She has known Niko long enough. Obviously, in his enthusiasm, he forgot that for once she didn't want to talk about bees. She doesn't want to spoil his joy and asks if he isn't hungry too: "Yes, very hungry, Gudrun, let's go to Sachsenhausen, Rippchen mit Kraut (Frankfurt pork cutlet with sauerkraut) and cider, we deserve that after such an important observation!" And with that, the joint visit to the zoo ends again with bees! Gudrun is a bit sad, even hungrier and the "Ebbelwoi" (cider) lifts the mood again. In the end the day is also successful for her. And Niko's new observation will certainly help him!

Niko resolves to precisely record the behavior of queen bees during egg laying. To do this, he spends days building a highly complicated observation hive in the Institute's basement. Obviously, he enjoys carpentry. He continues to be on fire for his topic and develops many ideas. The fact that so many researchers and beekeepers have thought and experimented in vain about the queen's ability to distinguish between worker and drone cells does not seem to depress or impress him. On the contrary, Niko takes great pleasure in criticizing and picking apart all the hypotheses described so far. Gudrun puzzles: Where does Niko get the conviction that he can solve this question better than the many people before him? Nobody has ever tried their hand at her topic. It is much safer that she will break new ground, especially since she can use the expertise of Professor Ruttner and Professor Pfleiderer.

The spring of 1966 comes late, with many cold snaps. Gudrun's queen production cannot yet start and Niko's practical experimental work in Oberursel is also progressing slowly. The wood preservative he used to paint his painstakingly built observation box to protect it from rain contained an insecticide: The bees are dying! He uses various solvents to try to wash off the impregnation. In vain. He loses valuable weeks until he finally realizes that he can't go on with this hive. The canister with the paint said that the painted wood is well protected against insect damage. He should have read the fine print more closely. He is particularly annoyed by Mr. Blotz's comments at breakfast, who can hardly contain himself with schadenfreude (pleasure in the misfortune of others). Niko is not spared—he has to build a new observation hive. Professor Ruttner preaches patience. How many months does such a bee season have in Germany?

Gudrun's work also falters. On the advice of Professor Ruttner, she initially concentrated only on queen breeding. She is supposed to deep freeze the hatched queens and only chemically process them when many have been collected. At the end of the bee season, there are 40 queens. With luck and skill, it will have a maximum of 40 microliters for chemical analysis. According to Professor Pfleiderer, far too little. Nevertheless, she is looking for all the analysis devices together. She finally wants to start the first experiments. When she thaws the frozen queens, the spermathecae burst like glass bottles filled with frozen water. Nobody had thought of that!! The fluid is lost in the abdominal cavities of the thawed queens. All efforts were in vain! She reports her failure to Professor Ruttner, who had recommended freezing:

"One of us could have thought of it. We knew that these damn bubbles were filled to the brim with fluid."

Professor Ruttner's comment is not very comforting: "When things are done for the first time, nothing else is research, then such failures are inevitable!"

The frustration runs deep among the young researchers: a whole bee season for nothing. What did Gudrun explain at length about bee research at the beginning? You can't do experiments with bees in winter—how right she was! And what now?

Niko and Gudrun discuss their miserable situation and come to the conclusion that they need to move to a warmer climate zone after the failed German summer season. They seek advice from Professor Ruttner, who likes the idea:

"Where do you want to go? As President of the Biological Commission of Apimondia, the international association of beekeepers, I have quite good relations with many countries. Maybe you should work in North Africa or the Middle East?"

Gudrun and Niko are happily surprised by Professor Ruttner's spontaneous support. Climate data and travel guides are studied. Niko leans towards Egypt, but Gudrun's decision falls on Israel: "For me, Israel is more than just a good honey bee country. Our parents' generation killed millions of Jews, and many who were able to flee now live in Israel in a whole new democratic state. At school, I learned about the kibbutz and how everyone is equal there: women and men with equal rights as the basis of coexistence. That's a much better environment for research than Egypt, where the role of women is still more than dubious, isn't it?"

"You're right. This is a strong argument for Israel. Research needs a modern social order."

Professor Ruttner immediately writes to the Israeli Ministry of Agriculture and asks for an official invitation. The officer responsible for beekeeping, Mr. Bar-Cohen, gets back to him immediately and asks for Niko's and Gudrun's personal data, CVs and a short description of the research projects. Two weeks later, Professor Ruttner comes into the lab, beaming with joy: "Just look. A letter from Israel! They are looking forward to your visit!" The two are over the moon: There is actually an official invitation letter from Tel Aviv in front of them! From October to May, they can work at the bee station in Zrifin. They are even allowed to live there on the ward free of charge. Job and accommodation are therefore secured. They also have to eat and drink in Germany, so no extra costs. That leaves the travel costs. How much will it cost to travel to Israel? They expect an amount between 2,000 and 3,000 marks. They don't have that much money. Their joy gives way to the question of how they could earn enough money in a short time. "You know what," says the professor, "I have good connections to the German Research Foundation (DFG). I will inquire whether there is a possibility to receive a travel grant for you."

A few days pass with waiting. Finally, Professor Ruttner summons Gudrun and Niko to his study. He urgently needs to talk to them. Gudrun is worried, because he makes a serious face. This is unusual. Professor Ruttner usually discusses everything with them at breakfast or at their workplace. Especially when it comes to good news, the breakfast room is the place of choice. Everyone should know that. The news must be bad. Now even Niko becomes nervous, despite his usual optimism. "There's probably no money and nothing works," says Gudrun.

"Let's just wait and hear what Professor Ruttner wants." Very anxious and with bare nerves, they then stand in front of Professor Ruttner's desk. This time he also seems to have lost some of its Austrian lightness: "So, eh, sit down." He seems insecure and tense: "As a matter of principle, I don't interfere in the private affairs of others. You know this, and you haven't heard a comment from me about your relationship until now. Yesterday on the phone ..." Professor Ruttner looks out of the window "... asked the DFG officer whether you, eh, well, are married." Professor Ruttner has regained his composure and looks at them: "A joint trip abroad is not possible if the relationship between you has not been officially clarified. To make a long story short, the lady from the DFG thinks that only the travel expenses of a married couple can be supported."

Professor Ruttner looks at them with raised eyebrows.

Gudrun laughs out loud: "I thought it would be something bad and the DFG wouldn't pay at all." Niko looks at Gudrun in surprise and then says loudly and very determinedly: "We'll get married. Agreed, Gudrun?" "Well, of course. Do we want to go to Israel or not?" Outside it is pouring rain. "Do we want to compensate for this lousy bee season or not?" Professor Ruttner is obviously relieved and laughs along: "I imagined it would be more complicated! There are couples where something like a wedding has to be planned years in advance."

Back in the lab, Niko asks Gudrun: "Blackmail by the "Deutsche Forschungsgemeinschaft"? Gudrun looks at Niko with a laugh: "Israel is a very important prize for us. We want, yes, we have to go there! By the way, we love each other, don't we?" Niko reacts irritated: "Dear Gudrun, what does our love have to do with the regulations of the DFG?" "Niko, you're right, nothing at all. If you or we want to go to Israel, we should get married as soon as possible!" Gudrun swings to Niko's line.

How does that work? Gudrun and Niko have no idea. Saying "yes" to the Professor is by no means enough. That much they know. There are important rules that must be observed. Now Gudrun's parents come into play. Gudrun calls her mother and she is delighted that Gudrun is finally officially "under the hood." Gudrun is annoyed by her pathetic demeanor. The parents rave about a glittering party, wedding in white and much more. Gudrun makes it short: "I'm not pregnant, but we still have to get married as soon as possible. Preferably this week."

Her mother is outraged by this rude approach: "Just don't think it's going to happen so quickly. It takes at least four to six weeks."

When Gudrun explains the circumstances to her in detail she becomes more merciful and arranges the formalities at the registry office in Wilhelmshaven. The family record book including all documents was delivered from Glückstadt for her sister's wedding four years ago.

In Bad Salzuflen it is more difficult, birth certificates have to be found and delivered to the registry office. "Do you have them?" asks Niko on the phone. "Of course not! You were born in the middle of the war." The administration searches for the birth certificates in Niko's birthplace and finds what they are looking for in Berlin. It takes time for certified copies to arrive in Bad Salzuflen. Then the marriage must be displayed for two weeks before the ceremony can take place. What an unnecessary delay, Gudrun thinks. Finally, the date is set. Niko and Gudrun learn they need two witnesses, and ask two friends from Freiburg. Uff and Michael, who have time and are happy to be part of such an important event! Nobody really takes the procedure seriously, except for Gudrun's and Niko's parents, who are deeply disappointed that the young people do not appreciate this ritual and "important step" in life. There are only two weeks left for the wedding preparations because of the work with the bees. That's plenty of time. Niko digs out his old suit, his dance lesson outfit which still fits. Gudrun needs a dark costume. Frankfurt has a lot of choices. It won't work at Woolworth's, but it will at C&A. So far, so good. Everyone at the Institute has something to tell. Wedding: The "most beautiful day in the life" of many brides turns out to be a first-class opportunity for "bad luck, mishaps and intrigues." Gudrun and Niko listen intently to these descriptions. Their decision to keep the "procedure" low-key is further strengthened. Gudrun is not completely immune to well-intentioned advice. The question, one day before the wedding trip to Wilhelmshaven, is about the need for indispensable white gloves along with a matching white handbag, embarrasses Gudrun: "Niko, what has to be, has to be. I need white gloves, and if the money lasts, a chic handbag! Marriage to me must be worth it to you?"

"Hurry up to Horex, my future wife. As a sign of my appreciation, I put a pair of elegant white shoes on top." At the Kaufhof near Hauptwache in Frankfurt, these wedding wishes become reality.

On Thursday evening, Michael and Uff, the groomsmen, arrive in Frankfurt with Michael's VW Beetle as agreed. Without eating together in the Kiosk around the corner, the Freiburg friends cannot continue. They got off later than hoped. Shortly before Dortmund, the motorway is closed. Nothing works for hours. Only after midnight does it continue relatively quickly to Cloppenburg. However, Michael's Beetle is not one of the fastest vehicles. Just an used car with a not quite powerful engine. In addition, there are 100 kilometers of country road behind Cloppenburg and Michael hadn't really taken that into account with the planned travel time. They don't arrive in Wilhelmshaven until Friday morning, completely exhausted, and Gudrun's parents don't allow Niko to spend the short night in their house. That's not right! Where would we end up? He has to sleep with the groomsmen in the house of the Association of Christian Young Men and must also change there. Nothing works before the wedding! Michael and Uff are also booked there for the nights of Friday and Saturday. They finally arrive just in time to wash and dress up for the ceremony. Everything is going well, just like that.

Punctually at 10 a.m. on Friday, Gudrun and Niko's wedding ceremony takes place in the town hall in Wilhelmshaven. Gudrun says "Yes", so does Niko, and that's it (Fig. 9). They quickly go up to the town hall tower, where there is a wonderful view over Wilhelmshaven and the Jade Bay. On the way down, Gudrun has to laugh: "This is probably the shortest honeymoon trip ever in Wilhelmshaven."

Niko gives her a kiss as they reach the bottom: "That was an unforgettable honeymoon!"

The wedding party listens and laughs, but some are piqued by such mockery of traditional rituals. The wedding celebration, to which Niko's parents and some siblings travel, is not a big celebration. The bride, groom and groomsmen are dog-tired. Malicious tongues claim that they fell asleep on the sofa, which Gudrun and Niko will resolutely deny for the rest of their lives. For the wedding night, Gudrun's parents have booked a room for them in the best hotel in Wilhelmshaven.

Just two days after the wedding, they are back at the Institute in Oberursel. In front of his observation hive, Niko experiences a nasty surprise: the queen is nowhere to be found. The next day it has become a certainty that she is lost. A bitter loss: The planned experiments cannot take place! Niko quickly gets a new bee colony and only a few days later he begins to observe the new queen. That doesn't change the fact that Gudrun often has to hear later: "My queen died because of our wedding!"

Fig. 9. Wedding: Gudrun with new white gloves, white handbag and white "high heels", Niko in a suit from high school, right and left the groomsmen.

Fifth chapter

Gudrun and Niko travel to Marseille with the Horex

Although the German Research Foundation (DFG) gave the green light for the trip to Israel, the budget for Gudrun and Niko is quite manageable. Apart from the direct travel costs, nothing else is covered. However, the two are now experienced and highly professional in traveling economically. You pack your bags, hop on the Horex and go wherever you want. By motorcycle to Israel? Niko takes the somewhat disheveled road atlas from the shelf. Further back in the European overview section on a scale of 1:150,000, Niko traces the route with his finger. He has to leaf through a lot. The map does not reach through Austria, Yugoslavia, Greece, Turkey and beyond.

"Hmm, long distance. Exciting. We should get road maps of the missing routes."

Gudrun shakes her head: "No, that may be exciting, but it's just too far. Also, it doesn't work. The Syrians won't let us through if they realize that we want to go to Israel. The Israelis will certainly not let us in if we come from Syria. They are all hostile to each other and overcomplicated. We have to find another way."

As a girl from the coast, Gudrun is aware of sea routes and studies the map. She is looking for ferry connections. "Where are you looking on the map? We have a motorcycle." "Wait and see." She finds dotted lines that lead from Genoa through the Mediterranean. "Genoa-Haifa. Bingo. Why don't we go by boat?"

She continues to search. There is also a ferry connection from Marseille to Haifa.

"We don't go via Genoa, but via Marseille! We'll save ourselves from traveling through the Alps." Gudrun is enthusiastic: "It wasn't a joy of traveling last time to Lunz through the mountains. When do we leave? In October? There is snow on the passes if we are unlucky." Niko admits defeat. The idea is brilliant, but Professor Ruttner comes from Austria and knows less about sea routes than Niko.

"Do you think we can convince Professor Ruttner? I have the impression that he finds motorcycling unsound." "That can be true. He makes pointed remarks about the Horex from time to time. That worked out after Lunz and there is no reasonable alternative to our plan. If we fly, we have to rent a car in Israel. That will be far too expensive for eight months. And why do we have the Horex?" Gudrun is not really sure, but she doesn't let it show. Niko is convinced and puts the atlas back on the shelf. Everything is fine! "No, please!" the Professor exclaims when they present their plan to him. "You don't want to go to Israel on a motorbike, do you?"

The two are on fire. Their arguments and especially their enthusiasm impress him. He has respect for students who know what they want, as long as his own goals are also implemented. Professor Ruttner quickly realizes that a lot of money is being saved. Flying is expensive. Almost 20 years after the Holocaust, there is still no direct flight connection from Germany to Israel. You have to fly from Frankfurt via Paris or Amsterdam. Or purchase tickets for the train to Marseille? The motorcycle is definitely cheaper! The DFG's funds could also be used elsewhere in the Institute's research budget.

"Hmm, you've thought it through. A clever plan. By motorbike! I could have guessed it. I hope the Israelis are not too frightened when two motorcycle youngsters invade their country." Smiling and shaking his head, he agrees. As they leave, they still hear Professor Ruttner calling to the secretary in the adjoining office: "Mrs. Hofmann, can you ask the travel agency for a Marseille-Haifa ship passage for two people and a motorcycle to Haifa. Return after 8 months. No, not for me, what are you thinking! Yes, to Haifa for Gudrun and Niko."

They received application forms for their visas from the Israeli embassy in Bonn. Since the date of entry is now fixed, they can now be filled out. Countless pages of small print. Thankfully, these sections are only relevant for applicants with a pre-1940 date of birth. They apparently want to be quite sure not to let any old Nazis into Israel. For younger people, the application is quickly filled out: passport numbers, date of entry, duration of stay, purpose of the trip, contact in Israel, and that's it. The departure is in four weeks. The postal service seems too risky. So off to Bad Godesberg. The motorway to Bonn is well developed. At Königswinter, they take the ferry across the Rhine, which drops them off directly in Bad Godesberg at the road Rheinallee. There, one embassy follows the next. In the quiet, tranquil villa district, their motorcycle is alone on the street and too loud for the villa district upscale area. There is not much activity when they unpack their documents in the counter room. They are the only ones. A lady takes their application and passports. Gudrun is excited: "We have traveled all the way from Frankfurt because we want to leave in four weeks." The lady nods understandingly, puts on her reading glasses and carefully reads the documents. "Ah, you're doing research on bees. Very unusual." The lady eyes the two in their motorcycle gear. She smiles as she continues to check the documents: "You certainly don't plan to travel to Israel on a motorcycle, do you?" Gudrun doesn't understand. "Yes, of course. How else," she blurts out.

The employee looks at her in surprise over her rimless reading glasses. Then she laughs: "Oh, yes, how else do you get to Israel?" She goes into the next room, where a colleague is apparently trying to get the visas and the stamps in the passports. They hear loud laughter, "bees", "motorcycle". The lady comes back and hands over the passports, still visibly amused: "Well, then I wish you a good trip. Don't get stung, and drive carefully!" Gudrun and Niko thank her, they don't quite understand why their request seems to be so funny. That's important research. The motorcycle ride is well planned. Why don't they understand? Well, the main thing is finished. They hop on the Horex to return to Frankfurt.

It takes a lot of time before they hold the ship boarding tickets in their hands. The Oberursel travel agency has never been fast. Only after two endless weeks and several telephones calls the documents for the passage from Marseille to Haifa arrive at the Institute. A ship of the Israeli line Zim called "Moledet" will transport the two together with the Horex. Professor Ruttner hands over the tickets in his office with his characteristic pathos and beams all over his face:

"Please, the tickets. You have to prepare well. It will not be easy without the Institute and our help in a foreign country. I am sure that you will succeed and both of you will do everything not to disappoint my trust and that of the DFG. Please don't forget the weekly report on the progress of the work!" Gudrun and Niko thank him politely. They are mainly relieved that the wait is over. The disappointments about last summer's unsuccessful work run deep. Both absolutely need initial research successes. This winter in Israel will bring the necessary results. In the climate there, everything has to work out with the bees.

Now it's time to make the final travel preparations. A sabbatical is quickly registered at the University. The most important thing is to check the Horex again to see if it is fit for the long journey. There is an annoying problem with the transmission oil. The drain plug has been leaking and Niko has to change the seal. When the machine is ready to travel. Gudrun is relieved. Niko enters the laboratory satisfied, albeit with oily hands. "All right or do we still need spare parts?"

"Oh, with my combination pliers, wrenches and a binding wire, I can repair most of it myself. In Israel there are certainly motorcyclists and workshops that can help with major breakdowns."

The packing list has become a travel routine for them: two panniers and a backpack are more than enough for the eight months. Gudrun insists vainly to pack skirt and blouse: "What if we receive an invitation to the ministry? I can't show up in my motorcycle jeans and the old canvas jacket, can I?"

The really important things go into the breast pouch: passport, letter of invitation, ship tickets, the American Express dollar traveler's checks. Gudrun has also changed French francs and Israeli lira so that they don't starve to death during the outbound journey.

On the day of departure, they go back to the Institute to say goodbye. The entire staff of the Institute stands in line in the courtyard as the two set off on the fully loaded motorcycle. Mrs. Hofmann and Mrs. Agnes shed a tear, master beekeeper Blotz does not. Professor Ruttner has an important contribution to the departure:

"I hope you'll taken everything you need. And please don't forget the reports!"

Niko starts and accelerates the Horex (Fig.10). Gudrun rolls her eyes behind her motorcycle goggles, holds on to Niko with one hand and waves politely behind her with the other. They don't see the Oberursel team waving at them to the corner.

The weather couldn't be better: sunshine and a few clouds. Despite the heavy side pockets, the Horex is solid on the highway. Niko initially planned to ride moderately, but the beautiful weather and the powerful machine increase their speed. Gudrun acts as cruise control. Her hands claw at Niko's chest when he goes too fast.

The first stage takes them to their old friends from their time in Freiburg, Maria and Herbert. Even with the Gudrun's cruise control, they arrive in Bollschweil an hour earlier than planned. In the remote area, the children have heard the Horex from afar. Suse and Julius run towards them with loud shouts of welcome.

"What's that? A motorcycle—what happened to the beautiful Heinkel scooter?"

Gudrun has to dismount to cuddle both children. Then they have inspect the Horex. They are now big enough to climb onto the motorcycle themselves, Suse in the front on the tank, Julius on the back seat. They cling to Niko: "Come on, go!"

As in the past, he lets the motorcycle roar vigorously several times. Being loud is everything. With a rattle and a few deliberately banging misfires, he slowly but unmistakably drives a lap of honour through the village with the two children. When they return,

Fig. 10. Starting with the Horex to Marseille.

Herbert and Maria are standing in front of the front door, shaking their heads and laughing. Herbert wears slightly horn-rimmed glasses, which extend his prominent nose even further. Despite the new baby, Maria looks as if nothing has happened: sporty and naturally elegant. In the living room there is brewed coffee as in the old days. Julius on Gudrun's lap and Suse shares the chair with Niko, preventing any serious conversation, and their answers to the questions about why they are going to Marseille have to wait until both children are in bed. Niko even tells the two of them a bedtime story. Today it is one of the old motorcycle, which is finally allowed in the garage after the long, long distance.

Herbert has uncorked a bottle of Pinot Noir and in the living room they now report in detail about their work and failures of the past summer. Herbert and Maria can't really understand why experiments should be carried out on bees at all. Nature has fixed it, why do you have to investigate something that the bees apparently do well? Only when they mention the invitation from the Israeli Ministry of Agriculture does Herbert prick up his ears: "Germans officially invited to Israel? That's not possible! We have only just established diplomatic relations with Israel." Niko is surprised. He has not followed the special turmoil of diplomatic relations between the young Federal Republic of Germany and Israel. He wonders about such concerns and shakes his head. "I didn't notice anything about the political entanglements when I was preparing the trip. The only problem was the people from the travel agency, who had a terrible time organizing the tickets for the ferry. Otherwise, everything was very uncomplicated and relaxed: Professor Ruttner is President of the Biological Commission of Apimondia, which is the World Bee Organization. As a result, he somehow knows everyone who is important in bee research. He asked the head of the Israeli delegation, his name is Bar Cohen or something, to get the invitation for us. They were very excited: Great project, must be done, we are happy to participate. That's how they wrote it to us." Gudrun continues: "Everything is also well regulated financially. The travel costs to Israel are paid by the German Research Foundation. The DFG has probably only recently launched an Israel program for scholar exchange. This was only possible for us as a married couple. Otherwise, the DFG would probably have been guilty of matchmaking. That's why we had to get married quickly." Gudrun and Niko beam at each other and show their unengraved rings. Gudrun reports on the efficiency with which they did justice to all German wedding rites. "We even managed our honeymoon. Once up to the town hall tower in Wilhelmshaven with a view of the sea, of course—over the Jade Bay."

Maria and Herbert are impressed by the speedy wedding ritual, but are happy that they took a little more time. "The regulation is as follows: The DFG covers the travel costs and the Israelis provide the accommodation. Bingo. It's totally uncomplicated."

"There is also no scientific problem at all. Mr. Bar Cohen knows our program and was really enthusiastic." Maria has doubts: "Do you really think that you are welcome in Israel, after all that was done to Jews by Germans? There is certainly a lot of resentment. They first threw stones at the new German ambassador, or tomatoes? That was only last year." "That was two years ago," Herbert explains. "His name is Pauls, I think. Rolf Pauls. On the one hand, he was a Knight's Cross bearer and a member of the Wehrmacht, on the other hand, he was also active in the resistance against Hitler. Not an easy CV and certainly not an easy choice as Ambassador, especially not for Israel." Niko generously brushes away all the concerns: "Well, this may all be new and complicated for politics. We are researching honey bees! I have no Knight's Cross and Gudrun no Mother's Cross. These are old stories from back then. The ministry could have simply cancelled if we were unwanted! No, the Israelis find our research exciting and want to support it. They are also honey bee researchers. It doesn't matter which passport we have. For our experiments, Israel's climate data are very favorable. We can do our experiments there in the middle of winter. Who are we supposed to disturb with this? Besides, we are really not the perpetrators of that time. They know that too."

Neither Herbert nor Maria seem convinced and look at each other with raised eyebrows. Maria probes: "And what about Palestinian terror? Doesn't a bomb go off there every day?" "Oh, Maria, the danger of terrorism is hopelessly exaggerated. If there were no reporting media, any terror would be pointless. In Israel, many more people die from road traffic and heart attacks than from terrorist attacks. The ward where we work is also located in an absolutely safe area." Niko explains this so convincingly that he almost believes it himself. He secretly hopes that his spontaneous explanation is somehow true. That Maria doesn't ask where exactly the ward is if it's so safe there. He is lucky: a big gulp of the red wine allows a change of theme to the old Freiburg student scene. Herbert reports that the exmatriculation (expelling) of their old friend Jacky Gellert is now finally resolved: "She had sprayed 'Break the Power of the Full Professors' on the wall of the Mensa and even boasted about it publicly." Maria contradicts as usual: "Jacky didn't brag. What sense does that make if you don't publicly stand by your convictions. And the statement was always correct. The professors can do what they want. If a student's attitude is not to their liking, the next exam could be assessed accordingly."

It goes back and forth between Herbert and Maria like in the old days. Tonight, Gudrun and Niko are too tired to actively participate in such a discussion. Tomorrow they want to drive to Avignon. The emptied glasses remain empty and they go to the guest sofa.

The start the next morning is not without delays. The children urgently need to be driven to the kindergarten by motorcycle: "So that everyone can see that we really ride motorcycles!"

Another lap through Bollschweil. The Valley is shaking. After a short heartfelt farewell to Herbert and Maria, the journey continues. At the French border, everything is thoroughly inspected. The customs officers have a keen eye on suspects with contraband. Cigarettes? Butter? Spirits? The cars are bumper to bumper. Niko drives slender past the queue of cars and so they quickly come to the barrier. There is not much to smuggle with a motorcycle and the French border guard is more interested in the Horex than in duty-unpaid goods. He asks Niko something about "vitesse?" Gudrun fibbers something about 150 to 160 km/hr. The man is impressed, nods appreciatively. Gudrun fumbles in her breast pouch in search of the passports, he just waves them on, laughing. There are no motorways in France yet. On the other hand, the country road is often three-lane, with a dangerous middle fast lane. "If you're going to do it head-on, then do it right," Gudrun thinks. There are certainly more people dying on the streets here than in attacks in Israel. The landscape is beautiful, but the route to the Rhône valley is mountainous and winding. They don't pass Lyon until the early afternoon, after which they treat themselves to a longer break with baguettes, salami, chocolate and French pastries. The tested and proven travel catering! Neither cola nor chocolate help: After the rest, all limbs still hurt. It is only with difficulty that they swing back onto the machine. Behind Valence at a gas stop, they look for a hotel, wanting one as soon as possible. Continuing south on the N7, behind Montélimar, they finally discover a parking lot with many trucks, behind them the neon sign of a hotel.

"Long-distance drivers know where to find cheap food and accommodation."

Gudrun struggles from the pillion seat. "LE PELICAN" is the neon lettering in capital letters above the round entrance door. The "LE" flickers irregularly.

"Let's see what the Pelican offers," Gudrun jokes.

The concierge at the counter welcomes them warmly. He examines their motorcycle gear, while Gudrun asks for a double room in perfect school French. He answers in German:

"No problem. There's even room for your motorcycle in the garage."

His Dutch accent is also unmistakable in German.

"Where is the journey going?"

"To Marseille, and then on to Israel."

"Huh? That's not possible."

Surprised, he lapses into Dutch: "Naar de Joden? They laten jullie in het Land? They're crazy! I've always known that."

Gudrun and Niko are perplexed that they are meeting with the same scepticism as in Bollschweil.

Gudrun explains: "We do research on honey bees and don't think we'll have any particular problems there."

"No problems?" The concierge laughs, shaking his head: "Jongens, you just killed them all, didn't you? Or? And now you want to go there of all places? You have a German license plate. That's not good advertising, hey. Not here and certainly not at all there. That's why the engine is in the garage. To Israel? I don't want to go with you." Niko is too tired and doesn't feel like reopening the discussion with Herbert and Maria here. "Well, unfortunately we don't have a choice now. We are officially invited by the Israeli Ministry of Agriculture and now we have to go."

The concierge is not reassured and warns them again when he gives them the key: "Just watch out. As recently as last year in the summer, German cars have a flat tire here in the parking lot from time to time, French cars never. That's the only reason why I converted de oude Schuur's, the old barn into a garage. The war has only been over for 20 years, but there are still a lot of bad memories."

Gudrun and Niko ignore the dramatic warnings. How often had they driven from Freiburg to Alsace and never been met with hostility. They never had any problems. On the contrary, they had always been warmly welcomed. And now this? As they go up the stairs to the first floor, Niko sums up: "Such an over-the-top nutcase. I think he's just seen too many war movies." Next morning Niko misses ham and cheese for breakfast. After a large café au lait and two warm croissants to dip in, they continue south on the N7.

Sixth chapter

Gudrun and Niko dine French

They arrive in Marseille around noon, two days before the departure of the Moledet. The traffic confusion of the metropolis does not bother them. Many traffic situations here are negotiated spontaneously rather than following a binding set of rules. They adapt to the new conditions. It is not easy to obey a red light while waiting politely until it turns green. Especially when there is loud honking.

Around the Place de Joliette at the ferry port, accommodation is far too expensive. It is only in the evening that they find what they are looking for in the district at the old harbour. The "Sable d'Or" advertises acceptable prices. The plaster is crumbling, the front door squeaks and a worn red rug shows the way to the reception. If there were ever good days in this hostel, they are long gone. The concierge greets them politely, pulls out the registration form and asks for their passports. "Aah, allemand? Vous êtes mariés?"

The question of whether they are married catches them unexpectedly. This is not evident from the passports. On Gudrun's passport there is a note on the last page in German that she was married two months ago and is now called Gudrun Koeniger. Her references to the small print there and the somewhat nervous explanations are noted with a serene smile: "Pas de problème"—and the key is discreetly pushed over the counter against advance payment: "Chambre 23". The concierge points to a narrow wooden staircase that leads steeply up to the second floor. Their room: A "Grand Lit", a chair and a closet without doors. It smells of cold ashtrays and stale air. Gudrun opens the window. No sea views! She looks at a meter-high mountain of garbage in the backyard. A stench of rotten fish rises to her nose. She quickly closes the window again, they look at each other: "Two nights, it's going to be okay!" Later they realize that most of the rooms in this hotel are rented by the hour. Here, no one cares about sea views. Loud price negotiations and the noises from the neighboring rooms keep them awake for a long time. Fortunately, the feared raid by the vice squad does not materialize.

Gudrun and Niko enjoy the day off in Marseille. A crossing to Monte Cristo Island and an extended snorkeling tour around the rocks are good for both of them. In the evening, they decide to go out for a real French meal instead of the usual combo. On the quay of the old harbor, one fish restaurant follows the next. Even with the first one, reading the menu is exhilarating and triggers a saliva reflex. And the prices are astronomical! So off to the next restaurant. But here, too, the prices on the posted menu card do not fit their small travel budget at all. Twenty restaurants later at the end of the "Quai du Port", both are unnerved, hungry and quite perplexed. Another baguette? The mood sinks. A man, perhaps in his early 50s, pale and gaunt with a wrinkled pre-war double-breasted suit, oversized nickel glasses, beret, casual and yet somehow elegant, approaches Gudrun and Niko. The twirled moustache underlines his appearance. More cliché is not possible. He addresses Niko:

"Can I help you?"

Oh, French, eh, like "aider". What's that called? Normally, Gudrun would come into play now. To Niko's miserable French, they immediately get an answer in perfect German. Slightly North German colored: "Oh, you're from Germany. I'm Christophe!" and stretches out his hand. They are completely perplexed and take it a little hesitantly "Gudrun, Niko." "Are you looking for a restaurant?" "Eh, yes, but why do you speak such excellent German?" "A long history. I was a prisoner of war on a large farm in a small village near Hanover for five years. Yes, five long years!"

He pauses and looks briefly over the harbor in his thoughts."Maybe I can help with your problem. Here a suggestion." A short smile. "I'm a chef and I own a small restaurant on Rue Refuge. Today is our day of rest and we are actually closed. I would like to invite you to a small and modest dinner and you can tell me what life is like in Germany today. Come to 38 Rue du Refuge later at nine o'clock, and just knock loudly."

Christophe laughs and disappears across the street and leaves them standing a little puzzled. They look at each other in silence. A former prisoner of war who, despite everything, has good memories of Germany? Is there such a thing? Shouldn't he rather be one of the vengeful tire stabbers that the concierge at Le Pelican warned them about? In any case, the prospect of a free dinner is enticing. On the way to the hotel, their doubts catch up with them and they discuss the unusual invitation for a while. What did you think prompted this Christophe to spontaneously invite two Germans who were completely unknown to him? Has he had contradictory experiences in captivity and now wants to hear what happened to them after the war? Is the invitation possibly not meant seriously at all? Or are they even falling into a trap?

Gudrun decides: "It doesn't matter! In view of our travel budget, an adventure is always better than an expensive evening." Departing on time, and thus at least half an hour early for France, they arrive at the small restaurant. It is within walking distance of their hotel. However, it looks much better than her shabby hourly hotel. There is no crumbling plaster and the golden "Chez Christophe" above the entrance looks freshly polished. Before they knock, they take a quick look at the prices of the menu on display. They are amazed. Here everything is even more expensive than in the restaurants at the harbor. "Chez Christophe" is apparently not just any restaurant. On the menu card they see a picture of the laughing Christophe in a chef's outfit with a high chef's hat, who proudly has a gold medal around his neck. Gudrun says dryly: "It seems that this is the good restaurant." She points to a red plaque with a star, underneath a small waving Michelin man. Niko knocks three times. A beaming smile greets them. The woman who smiles introduces herself as Christophe's wife Josette. Brown shoulder-length hair, a little makeup, lipstick and eye shadow. She asks them to the only table set. With white tablecloth, set for three people. Josette is not joining us for dinner? Niko and Gudrun are impressed by the cutlery and the many different glasses. They remember that they have no idea which glass goes with which drink. "My husband is still in the kitchen. He'll come right away." Josette disappears to the back. Perhaps she had different experiences with Germans during the war than Christophe? They wait, standing and somewhat lost at the set table in the middle of the empty restaurant, until finally Christophe comes in with a white apron and chef's hat, a big smile and a large steaming soup pot. Like in the photo outside the door. "Welcome. I am very happy that you have come." They shake hands. "Why don't you take a seat?" He puts his apron and chef's hat behind the counter: "I don't need that anymore. Now we can drink and eat. Here with us, we drink pastis before eating. This is an anise drink! I've got a Pernod here!" He gives a strong shot into one of the larger glasses and fills it with water. Gudrun and Niko are happy about the silvery color that is created. They toast: "Cheers!", "À votre santé" and take a big sip. The anise taste is unusual. Gudrun grimaces: "Oh, yes. Anise is probably very healthy?"

Christophe laughs and explains that the effect of this drink depends on the quantity. They start with the hot herb soup and soon – was it after the second or third glass of Pernod? – their cheeks become redder and the conversations livelier. Christophe says that he was taken prisoner by the Germans in the first weeks of the war.

"That was not very glorious for me." Initially, he was interned in a prison camp near Hanover. He was assigned to farm work. Despite the hard work, there was always little food in the camp. "We were very hungry! It was terrible. In the evening, after working in the fields, I was locked up with the other prisoners in well-secured barracks. At some point the guards were dismissed and no one knew why. I was relocated to a farmhouse. There I was given free board and lodging. The farmers were extremely friendly, not like the guards in the barracks. There were only women and a "grand dad" who spoke fluent French and gave me language lessons in German. Hence my modest knowledge of German!" Gudrun smiles mischievously and is sure that there were also French lessons for the young women. In the meantime, they were at the second course, a sumptuous bouillabaisse with mussels, squid and pieces of fish. Christophe suggests switching to red wine.

"Fish soup and Pernod, that doesn't fit at all!" Gudrun and Niko are happy about the large carafe that Christophe brings. Two liters of red wine, will the three of them still make it tonight? The spicy and unusually hot soup makes them thirsty and the tart red wine tastes much better to them than the Pernod. After the fish soup, Christophe gets the third course. To his great regret, he didn't get any more lobsters because of his late invitation. Instead, they have to be satisfied with beef tenderloin. It is served with eggplant slices, fried with thyme and other herbs of Provence in olive oil. As a reminder of Christophe's time in Germany, there is a large pan of fried potatoes:

"I often made breakfast there with eggs, bacon and fried potatoes."

Gudrun and Niko are overwhelmed: "It's much more than we can eat."

Christophe ignores such objections and raises his glass: "A votre santé, start first!"

The glass of red wine is quickly emptied. Christophe says that he often wishes he could hear real German and practice his German. When they ask why he doesn't go to Germany and visit the farmers, who were probably decent to him, they get an evasive answer: "You should let the past rest. It was a special time in the war, extraordinary circumstances. After the war, I was sent back to France by the British. Normal life returns and Marseille is now my life. I love Josette, we are doing well and one shouldn't run after old dreams."

Then Christophe wants to know what they are doing and why they are here. No, he is not interested in bees, he wants to know how they got married. When he hears about the nightly trip to Wilhelmshaven and the honeymoon in the town hall tower, he is stunned: "What will become of Germany if you give up your most important traditions?" "Many traditions in Germany have not proven themselves well. It might be helpful to keep your distance for the time being."

Christophe nods thoughtfully and understands Gudrun's objection "Yes, you may be right." They ask Christophe about his own wedding. The eyes sparkle:

"Oh, that was a big celebration with a mass and almost 100 guests, toute la famille. We danced, drank a lot and ate even more!"

Towards the end, Christophe's story sounds not only like savoir-vivre, how to live. Something like an unforgotten love in Germany sounds through his words. A topic that they feel this evening, but do not address. Maybe at the next meal. A chocolate mousse, excellently matched with coffee, ends the story anyway. Afterwards, Christophe puts an opulent cheese platter on the table. He smiles and quotes the saying he learned on the farm: "Cheese closes the stomach!"

They take this as an indication that they have come to the end. They miss the coffee and the cognac and a perhaps another outstanding story about fraternization in northern Germany during the war. In any case, it is well after midnight, and despite the lively red wine mood: tomorrow, no: this morning the Moledet is sailing.

Seventh chapter

Gudrun and Niko embark and meet Dwora and Fritz

There she lies, the Moledet (Fig.11). Not a normal ferry where the cars just drive on deck. A loading crane has to heave each vehicle on board through a loading hatch. Gudrun and Niko arrive just in time for embarkation, tired and with a slight headache. The Horex is the last vehicle in the queue at the loading crane, after which the barrier is folded down. Perhaps the slight residual alcohol also offers advantages: They laugh relaxed at the anxious faces of the passengers, who see their cars swinging back and forth at dizzying heights on the hook of the loading crane. Niko draws Gudrun's attention to an elderly gentleman who is following the trajectory of his vehicle with an outstretched arm. It lurches in the wind more than he would like. When the vehicle reaches the upper loading sill of the ship's hatch, what has to happen happens. The man turns away and covers his eyes with his hands. He doesn't want to see that. Then nothing happens. The crane operator skillfully juggles the vehicle a few millimeters past the edge of the hatch and into the ship undamaged. He looks up at the crane operator, who casually raises his thumb in his pulpit: I am good! The gentleman has to wipe the sweat from his forehead with a handkerchief and goes on board, shaking his head.

Now the Horex, the only motorcycle, floats with an elegant curve on the rope of the crane and disappears into the Moledet. Gudrun is in a good mood and waves to the crane operator. He recognizes her as the owner by her helmet, waves back and presses the alarm siren. To his delight, it echoes piercingly through the entire harbor. People react alarmed, pause and look for the cause of the alarm. Soon the connection between the motorcycle, the crane operator and the waving Gudrun is established. A nice end to the work for the crane operator. And on board, everyone knows Gudrun is the young woman with the big motorcycle.

On the narrow stairs leading to the cabins on the lower deck, there is a dense crowd. Heavy suitcases and travel bags are hurriedly dragged back and forth. From the cars that have found space at the bottom of the ship, passengers are pushing up. Niko and Gudrun have to go from top to bottom. Why the hustle and bustle? The cabin seats are printed on the ticket. Why hurry? Their cabins, each for six people, women or men, are located directly next to the ship's engine. No porthole to the outside. It's hot, the air is stuffy, it smells of lubricating oil and sweat. The heavy marine diesel engine is hammering at idle next door. The communal washrooms are located at the end of the narrow hallway, behind the stairs to the next higher deck. When they divide the things out of the backpack, they now even think wistfully of the shared room in the dodgy Sable d'Or. On board the Moledet, France is over for the time being: Now they are in Israel.

Punctually at half past eleven, it's time to cast off. The Moledet slowly slides out of the harbour, past the cathedral of La Major and sails around the offshore Les Iles and soon reaches the open sea, course south-east. For lunch, Gudrun and Niko are seated by the steward at a table of six with four other passengers. These will be her table neighbors for the next few days. They recognize the owner of the VW Beetle. He sits to the left of Gudrun, is now much calmer and talks relaxed to a woman to his right. His gray hair is neatly parted, but still looks uncombed. His shirt is worn open and his trousers, which are not very attractively bulged at the knees, make a careless impression. Under his bushy eyebrows, he follows the events at the table inconspicuously with attentive, deep brown eyes. All in all, an impressive man. The woman next to him, probably his wife, has dark hair with a slight shade of blue in large curls. Very well maintained. The

Fig. 11. Moledet in the port of Marseille

large-volume necklace is accurately matched to the clothing. There is another younger couple sitting at the table. Gudrun and Niko first introduce themselves to the VW owners in English. The answer in Iwrith (Hebrew) is reserved and determined: "Shalom". Gudrun and Niko didn't expect everyone to speak English, but rather expected "Welcome". Well, then "Shalom."

When food comes, Gudrun realizes again that they have left France. "Oh, Christophe, why didn't we stay with you," she thinks.

The atmosphere at the table seems rather tense. Did they put their foot in their mouth right at the beginning? Don't they like motorcyclists? Gudrun and Niko try to appear as relaxed as possible. They talk in German, the other passengers at their table in Iwrith. The quality of the food continues to decline. At least the ice shouldn't have melted!

"Milk soup comes to breakfast and hot on the table," comments Niko.

Gudrun laughs and the others at the table laugh with her. Gudrun and Niko are perplexed, look at each other in surprise, and feel a little caught. They then say "Auf Wiedersehen" in a friendly way and withdraw.

In the afternoon, the VW table neighbors sit next to them on the sun deck. Gudrun and Niko greet them with a somewhat artificial "Shalom". The response in German is whether they are the owners of the big motorcycle?

"And you own the VW?"

They laugh: "Yes, we stood next to each other during loading!"

Then the couple formally introduces themselves: "Dwora and Fritz Kuhn from Pardess Hanna!"

They left Germany before the war and are now returning from their first European vacation. They had been to Switzerland and Austria.

"Wonderful mountains". They did not go to Germany.

"Why not to Germany?" asks Niko. "We lived in Freiburg for several years. The Kaiserstuhl and the Black Forest are also beautiful!"

Fritz Kuhn's face darkens. There is a short pause:

"Well, Niko, it's one of those things with Germany. You have to understand: If we meet older men there, they could be the murderers of our relatives and friends."

Niko and Gudrun are shocked by this emotionally statement. They look at each other in dismay. Asked the wrong question? In any case, a relatable argument that they had neither thought nor heard before. They remember the warnings of the concierge of Le Pelican. When asked about the dismissive atmosphere at the table, the Kuhns apologized. For them, too, the conversation in Iwrith was unfamiliar. Normally they speak German to each other. In the presence of the other Israelis at the table, they had to demonstrate to the two young "German Christians" that they were Israeli and used this language. Dwora explains the language problem to the two:

"In public, Israelis avoid the German language. It is considered by many to be politically incorrect. In our private lives, it's different with us. In family and with good friends, we usually speak German."

Gudrun and Niko are happy that they seem to have been accepted into this circle.

"Why do you come to Israel as Germans? Weren't you warned that this could be fraught with complications?"

They talk in detail about their important bee experiments. Dwora soon loses interest and looks out to sea. Fritz continues to listen, but cannot imagine that their decision to select Israel was primarily determined by climate data and honey bee literature. He can imagine their research topics even less.

"Why do you do something like that? Nature has arranged it for the bees to function. The main thing is that they pollinate and make honey. You can sell it. Sperm in the queens receptaculum seminis? Queens that measure cells? What's the point?"

Fritz is perplexed and regrets that the two have been given such worthless tasks for their dissertations. Despite their enthusiasm, Gudrun and Niko also realize that Dwora can no longer suppress a yawn, and they should change the subject. However, this does not detract from their mood. In the following conversations, they learn that Dwora comes from Berlin and Fritz from Beuthen. Gudren and Niko have never heard of the town Beuthen.

"At that time it was a border town to Poland. In Upper Silesia. The city is now called Bytom and is located in the middle of Poland."

They know Upper Silesia. Their school atlas said: "under Polish administration".

"My father had a textile shop there in Bahnhofstraße. The Nazis forbade people to shop with us and after attacks on the store we had to give that up in 1937. We left Beuthen. Many others have stayed." Fritz Kuhn takes a break. The voice is occupied. "Auschwitz was only 50 kilometers away."

They know very well what that means.

"Oh Fritz, leave the young people alone with your sad stories!"

Dwora says that her parents, as enthusiastic Zionists, emigrated to Palestine in the early 1930s, even before the seizure of power by Hitler. She had come to the country as a "young thing", as she puts it. She first worked as an ironer and then as an accountant. Later, she took a job as a server in a café in Pardess Hanna. Many soldiers of the British army were stationed there, and mercenaries from various countries, not only from those of the Commonwealth.

"At the beginning, the groups sat separately, each according to their origin. After a short time and the first glasses of wine, the tables were moved together, people drank together and sang songs in all languages. And the highlight was always the Horrah. This is an Israeli folk dance. We danced, danced and danced until the early morning. That was in the early 1930s. Fritz came to Pardess Hanna much later."

She smiles contentedly as she looks at Fritz: "It was love at first sight. We got married quickly."

Dwora laughs at the memory, and then continues more thoughtfully: "Financially, it wasn't easy. We borrowed money from friends and got a used washing machine. Our laundry did not run as we had hoped. The washing machine broke constantly and had to be repaired and the laundry was not finished in time. Then the customers stayed away."

Dwora interrupts herself: "Why am I telling these old stories? Surely you must be bored? When I see you both, and especially you, Gudrun, such a young, beautiful woman, not much older than 20, it reminds me of the past and of the start with Fritz. Yes, we didn't have it easy but it was a full and very happy time."

She looks at her husband with a smile.

"Today, Fritz has a good salary. He works as a sales manager for a large lemonade factory. Our first, wonderful holiday in Europe! We have achieved a lot. And forgive us for attacking you with our old memories!"

Gudrun notices how little they actually know about Israel and the history of the Jews in the Third Reich: "No, no. That's exciting for us. We know so little about Israeli history. Instead of memorizing the dates of some Greek battles of Alexander the Great, we should have learned much more about Israel at school."

"You don't know anything about Israel and its history?" asks Fritz, surprised.

Niko also shakes his head.

"I was there from the beginning. When I came to Israel, it was still called Palestine and was a British colony. After the First World War, the British had wrested the land from the old Ottoman Empire. They had no interest in Jews. They have only allowed a very limited number to enter the country. Many Jews still wanted to go to Palestine! Not only the Zionists. Where else were they supposed to go when the persecution of the Jews began in Europe? The defensive policy of the British remained in place even when the National Socialists mercilessly persecuted Jews throughout Europe and everyone knew about it. Even during the war. Even when the extermination camps had long been known to the British. Immigration therefore had to take place illegally and under the strictest secrecy. That was the birth of the Mossad. With the help of this secret service, escape boats were organized to bring Jewish people to Israel, the Ma'apilim. The British had therefore set up a naval blockade off the coast of Palestine. When English warships caught ships carrying Ma'apilim immigrants, the passengers were interned in large camps in Cyprus and the ships were confiscated. Only a few ships managed to reach the coast of Palestine often after adventurous journeys at night. After radio contact with the secret Jewish defense organization, the immigrants were usually disembarked in the middle of the night somewhere between Jaffa and Ashdod. At agreed light signals, the people were rowed closer to the beach in lifeboats. Before they reached dry ground the last meters were in the water and the surf was very dangerous.

In my first years in Palestine, I often helped to bring the wet newcomers to the prepared hiding places with my donkey cart. We had to be careful that the British didn't discover us. The coast was long and there were only a few posts at night. The land patrols were very dangerous. The coast belonged to us, the Jewish underground movement Haganah. When refugees were picked up by the British, they were imprisoned in the Atlit internment camp, south of Haifa. Officially, I worked for the British and brought water to the camp in a large tank with the donkey cart. Water in, Ma'apalim out in the empty tank!"

Fritz grins. "That was my job and at the same time a perfect camouflage for the Haganah. The British never caught me."

"Then you were a double agent. Wasn't that dangerous?" asks Niko full of admiration.

"Perhaps, but in what danger were the others at that time? Compared to the gas chamber, it was a bathing trip for me."

Niko is silent. Every word is too much now. He now understands that the British also had no interest in the story of the founding of Israel being included in the curricula of German schools.

Eighth chapter

Gudrun and Niko experience rejection, curiosity and also friendship

The next evening at sea, Gudrun and Niko unexpectedly find themselves in trouble again: an elderly Israeli couple tells them that they left Israel and tried in vain to start a business in Frankfurt. After three years, they were on their way back to Israel. Many fellow travelers demonstrate their disapproval very clearly:

"It's your own fault! How can you go to Germany as a Jew and then as an Israeli?"

Turning to Gudrun and Niko, an elderly lady asks: "Jews must not live in Germany again! What do you mean? Do you want Jews in Germany again?"

Gudrun reacts shocked: "Of course we want Jews and Jewish communities in Germany again! In Frankfurt, many Jews came back after the war and there is a lively Jewish community there. In addition, freedom of religion is enshrined in our constitution. It doesn't matter who or what you believe in. We, for example, do not belong to any church and therefore have no prejudices!"

Gudrun emphasizes that it is above all a personal decision where you go and where you want to live: "The constitution explicitly regulates that all Jews expelled by the Nazis can regain German citizenship if they want to."

Then she adds timidly: "However, who wants to have it voluntarily today?"

Gudrun earns a lot of approval from the bystanders, but she can't let it stand like this.

"I can't change my passport. All I can do is change something in Germany and make sure that the Nazis and the brown horror don't happen again! You can be sure that anyone who wants to come to Germany is welcome there."

Niko intervenes: "I see you don't believe Gudrun. Our generation has a duty to ensure that Germany does not abandon the path set by the Basic Law and that everything remains good in the long run!"

That doesn't catch on. When they talk about the invitation from the Israeli Ministry of Agriculture, it triggers open disapproval: "How can two Germans be officially invited to Israel only 20 years after the Holocaust and the war?"

The passengers, in general, show great interest in this "strange German Christian couple" with the motorcycle. Gudrun and Niko realize that their presence on board is also somehow important—probably for many. The passengers thought they were sure that they would not find any Germans on this Jewish ship. For most of them, Gudrun and Niko represent the first encounter with the post-war generation from Germany. They trigger clear discomfort in many, but also skeptical curiosity. The two are deeply insecure. Germans are not welcome in Israel. Yes, they cannot be welcomed against the background of all those who had to flee Europe or survived the terrible horrors of the Nazi.

Contacts with young Jews are less stressful. The sunshine and warm weather are not only attracting Gudrun and Niko to the pool at the stern of the ship. There they often meet passengers who belong to their age group. It doesn't take long for them to start talking to each other. English is the language that everyone somehow masters more or less. They talk about studying and student life in the different countries. The other students from other European countries seem to be more conservative and tradition-conscious than Gudrun and Niko. Studying there is very tight. Niko fails in his attempt to translate " Unter den Talaren der Muff von 1000 Jahren!" (Under the gowns sits the musty of a thousand years!) into English. The meaning is lost and the others apparently have no problem with professors in gowns. The professor is wearing a gown. So what?

Niko meets Pablo from Barcelona for a game of chess in the bar in the evening. Spectators quickly arrive, forming a dense circle around the chessboard and commenting on the individual moves. Pablo doesn't seem to mind. Niko, who is slowly but surely losing, reacts nervously. He doesn't like the many spectators at all. What can he do about it? Niko puts his queen in front of a pawn. Pablo offers him to take the move back. Niko refuses and gives up the game. The viewers do not agree with Niko's approach and make it clear: "Bad loser", "bad loser!" Niko thinks he understands that for some spectators it's not just about a game. Here a Spanish Jew wins and a German Christian loses. And then the German ends the beautiful performance prematurely and cowardly. Niko smiles sheepishly. He apologizes: "Headache!"

No one believes him. Niko goes to Gudrun, who is sitting with Dwora Kuhn further back by the pool. When he complains that he lost the chess game, Gudrun shows no sympathy: "You've played and you're obviously not as good as Pablo. Why are you complaining?"

And Dwora remarks mockingly: "What did you expect? Of course, everyone here is happy when a Goy loses so clearly."

Niko shakes his head: "Well, I'm glad that Goy didn't win." All three of them laugh, but one doesn't sound entirely honest.

The Kuhns seem to have taken Gudrun and Niko to their hearts. They prepare the two for life in Israel with many stories. Fritz asks what they want to do when they arrive in Haifa on Friday. They have a simple plan: "We'll first go to the ministry in Tel Aviv. There we get the information about where they want to accommodate us."

Fritz shakes his head fatherly and smiles! "On Friday, you will not be able to reach a ministry office before the end of opening hours."

"Then we'll take a hotel room and go there on Saturday!" Fritz laughs loudly: "Saturday is Shabbat, I've explained that to you. Just like Sunday for you. Everything is closed and nothing goes! According to the rules of the Orthodox Jews, driving on Shabbat is also prohibited. In areas with an Orthodox population, stones are thrown at moving cars. Your loud motorcycle would certainly be a target."

Luckily, he and Dwora have talked about the problem: "We cordially invite you to come to our home for the weekend! Pardess Hanna is a small town founded by German Jews on the coast between Haifa and Tel Aviv. It's on the way for you, and on Sunday morning you can drive comfortably to your ministry."

Gudrun and Niko are impressed that after six weeks of vacation in Europe, the Kuhns invite two complete strangers to their home. This leaves them speechless. They thank them warmly, but point out that they cannot accept before they have arrived. Maybe they will be met by someone officially in Haifa? But Gudrun and Niko are now sure of one thing: with Dwora and Fritz they have found support, no matter what difficulties may come. A reassuring prospect in view of the uncertainties that are looming for them for the next eight months.

Finally, the coast of Israel comes into view and everyone cavorts on deck to see one or the other well-known landmark. The quartermaster informs that the cabins must be vacated within 30 minutes. Once again, there is a lot of back and forth. Gudrun and Niko have brought their panniers and backpack with them and can continue to follow how they approach the first port facilities. The port of Haifa is characterized by sober functional architecture, surrounded by state-of-the-art loading technology. As soon as the jetties are lowered, Gudrun and Niko are called out over loudspeakers. "Mr. and Mrs. Koeniger" are to meet with Mr. Rabinski at the jetty.

At the jetty stands a small, corpulent gentleman with gleaming nickel glasses, a white shirt with open collar and neatly ironed dark trousers: Mr. Rabinski, who introduces himself as the head of department of the Ministry of Agriculture. Even without a civil servant's tie, everything radiates authority, correctness and competence.

After the numerous critical remarks on board and their crash course in Israeli history, Gudrun and Niko have become nervous. Will there now be endless customs and immigration harassment because of German passports? By no means. Mr. Rabinski greets them politely, with a personal handshake and perfect Oxford English. Instead of customs harassment, he leads them through the diplomatic entrance, where they pass through without being checked. He is greeted by the officers with a military hand signal. They also stand at attention in front of Gudrun and Niko in motorcycle gear.

At passport control, things are similarly fast. Rabinski leads the way and only the visas are briefly inspected. Finally, the Horex is brought to them by two officials who are more interested in the machine than in its young drivers.

"I have reserved a room for you in a quiet guesthouse in Haifa," says Rabinski. "You know, tomorrow is Shabbat and everything is closed. You are not expected in Tel Aviv until Sunday at the ministry."

Niko thanks Mr. Rabinski and then tells him about the invitation to Pardess Hanna. Mr. Rabinski is astonished and raises his eyebrows. "Oh, an unexpected change of plans? Let's see what we can do."

Niko understands his surprise, after the overwhelming experiences they had with most of the passengers on board the Moledet. How could two Germans make friends with Israelis so quickly on that short cruise? Mr. Rabinski has Fritz and Dwora Kuhn proclaimed. When they arrive, they seem worried. Has customs possibly found prohibited goods in the VW Beetle? Fritz has the same tense expression on his face as he did when the car was loaded in Marseille. Dwora is also a bit tight-lipped. The two calm down quickly and there is a longer conversation in Iwrith with Mr. Rabinski. Dwora in particular seems to be leading the way. Whatever she says, after a while everyone laughs heartily and Mr. Rabinski turns to Gudrun and Niko:

"You want to go to Pardess Hanna with the Kuhn family?" "Yes, of course." They answer synchronously—and all laugh again. Pardess Hanna with Kuhns is of course much better than an anonymous room somewhere in Haifa. Mr. Rabinski goes to customs with Fritz and ensures that the car is cleared faster.

"Niko, that would never have gone like that in Germany. No pig would have been interested in that in Germany. No civil servant would have taken such personal sensitivities into account. Here, it is evidently the intellect that regulates the people, and not the rule that dominates the intellect."

Niko agrees thoughtfully as he puts on his motorcycle helmet: "Here in Israel, we are really very far away from Germany."

Ninth chapter

Gudrun and Niko come to Pardess Hanna, find the Ministry of Agriculture in Tel Aviv and reach their house in Zrifin

The road winds through the mountains and hills around Haifa and then continues through a dune landscape along the coast. The barren sand elevations to the right and left are overgrown with dry scrub. Where are the bees? From their motorcycle, Gudrun and Niko look in vain for apiaries. They are disappointed. This is not how they imagined the country, "where milk and honey are supposed to flow." Then they turn inland from the coastal road. Dry semi-desert everywhere. Where are the bees?

In the evening they reach Pardess Hanna, a small settlement dominated by flat bungalows, surrounded by dry vegetation. A small flat house in the middle of a large, well-kept and colorful garden. Gudrun and Niko are fascinated by the highly diverse plant and flower oasis. Something like this in the middle of the desert? A Garden of Eden. And finally, they hear honey bees buzzing in the midst of a rich insect life. Dwora has no understanding for their biologist interests: "First coffee!" She disappears into the kitchen. They go into the house and Fritz leads them into the living room. That doesn't seem to suit Israel at all. Wall cupboard, sofa, coffee table, TV in the corner: the bookshelves are stocked with German classics. In the display case behind the record player there are records with the music of Mozart, Schubert and Bach.

"It's like at home with our parents," Gudrun blurts out.

Fritz reacts irritated: "After all these years in Israel, our living room looks like that of

49

your parents in Germany? It seems that our tastes are shaped in childhood. That's why we probably brought some pages from Germany to Israel! Hopefully not the worst?", Fritz philosophizes.

"Now enough with your wisdom, Fritz!" Dwora comes out of the kitchen with the coffee cups. And after a cup of goodnight coffee and a few slices of bread, everyone goes to bed. Gudrun and Niko even have their own bathroom in the guest room. What a luxury after the cramped cabins and shared toilets on board the Moledet. Despite—or as the Kuhns believe—because of the coffee at the late hour, they sleep immediately and through until late the next morning.

After breakfast, Fritz proudly shows them his house. He explains how the small building was expanded step by step. It was all created with the help and expertise of neighbors and friends: "I had no idea about the construction, and we certainly didn't have any money. We had many, many friends! Here, everyone helps each other. A good path to success."

Fritz leads Gudrun and Niko through the house, shows them a large, overgrown plot of land behind his garden, which also belongs to him. He wants to plant and cultivate seedlings of the wild lemon "Chrush Rash" there.

"This is the base on which the orange trees are grafted. Lately there has been a shortage and prices have risen significantly!"

Niko realizes how important this project is to Fritz.

"Well, I'll help with the planting." Fritz is happily surprised to have found an ally.

"Really? Well, I'm happy to take you at your word. I can really use all the help I can get."

Niko is in good spirits, but has concerns about his lack of horticultural expertise:

"I have to warn you: I have no experience in planting anything. I worked asphalting motorways in Germany during the semester breaks. So I can do rough tackling, but doesn't it take a lot of skill and expertise here to make the trees a success?"

"Well, then you know how a shovel works. That's a good basis," jokes Fritz. "I can show you what goes beyond that. I'm not completely ignorant in fruit growing and horticulture! After the Nazis removed me from high school in Beuten, a Catholic master gardener took me on as an apprentice and trained me for over a year in his nursery. At the time, he ignored all warnings about training a Jewish apprentice. I was very grateful to him and learned a lot from him. The time was too short for the journeyman's examination, because the Nazis then also forbade that. That's when I emigrated to Palestine. When I arrived here, it looked completely different. Everything was dry desert. Then came the water pipes and with the irrigation the property changed. I had learned the most important basics in dealing with plants. That's how I came to the garden—and believe me," he rubs his hands: "Now it's no problem at all to successfully grow the lemon seedlings, and certainly not to get a good price for it."

On Sunday, Gudrun and Niko say goodbye to Dwora and Fritz after a hearty breakfast. They promise to come back in two weeks.

According to the map, it is less than 60 kilometers to Tel Aviv via National Road 2. In good spirits, Niko curves on the country road through the maquis landscape until the Mediterranean Sea comes into view on the horizon. There he turns south, onto the four-lane coastal road to Tel Aviv, which still has a wide gravel track for the donkey carts on both sides. It's going well—and then taking longer than they thought. The outskirts of Tel Aviv stretch and there is no city in sight. When will the center, where the ministries are located, finally come? After half an hour on site, the building density decreases again. Soon they will be driving through orange plantations again. They have obviously missed the right exit; the city is behind them. Niko makes a U-turn and drives back again. The only slightly taller house they have seen is now used as a landmark. The Shalom Tower, the only high-rise building in the city! And now?

Apparently, this is the center. At least there is a lot of traffic. They have to ask for directions. Luckily, Gudrun has put a piece of paper with the address in the pocket of her anorak. However, she has reservations about addressing passers-by directly because she does not know Iwrith. Due to the many different languages that immigrants have brought into the country and the former British Mandate territory, they do not seem to be the only ones with very limited knowledge of Iwrith. They also share their Hebrew illiteracy with many older immigrants. Many people who she asks in English answer in German or Yiddish. After several stops in the question, they arrive at the address given in the district of Hakirya. Flat single-story barracks everywhere! They had imagined ministerial buildings differently. The complex looks like a quickly cobbled together complex of emergency shelters, with the flair of an army camp from the Second World War. Later they will learn that several buildings are much older and were built by the Sarona community of the German Templar sect before the turn of the century. After the war, the British set up an army camp there, which was taken over by the Israeli government after their withdrawal. Apparently, the ministries were still housed in army barracks and the Israeli state had other priorities than erecting magnificent representative buildings for the administration. What a contrast to the ostentatious ministerial concrete castles in Bonn!

As soon as they drive into the parking lot with the Horex, a white-haired older gentleman approaches them. Obviously, Mr. Rabinski had warned him in advance and informed him about the two young people with the big motorcycle. In English, he introduced himself as Mr. Bar Cohen. So, this is the gentleman from whom the letter of invitation came! They thank him for the friendly correspondence and the important document that made their research project possible in the first place. Bar Cohen remains reserved. Their words of thanks seem to be uncomfortable for him. He politely invites them to the barracks. It almost seems embarrassing to him that they are now sitting in his office. It is as barren on the inside as the building is on the outside. Slightly dusty metal shelves are filled to the ceiling with files. In the middle of the room is a large, heavily used green-gray steel desk. The dark green linoleum work surface, bordered with a somewhat scuffed chrome strip, is covered with many documents and files. Above everything is an old black telephone that is ringing as they enter the room. Bar Cohen points to the two visitors' chairs and says: "Slicha." Sorry—one of the few words Niko picked up in Iwrith. However, the telephone conversation is not conducted in Iwrith, to their surprise in a Slavic language. It could be Polish. After hanging up, Bar Cohen turns back to them: "I talked to Mrs. Alpern. She runs our beekeeping station in Zrifin and looks after about 50 bee colonies there. She is looking forward to your arrival and asks how many colonies you will need for your experimental program. Unfortunately, all bee colonies of the station are involved in a large research program, so we need additional colonies for your experiments. I would then order the colonies from a reliable beekeeper. We want to be on the safe side and not buy any bee colonies with which we bring diseases or nasty stinging bees to the station. A colony costs them about 30 pounds, depending on how strong it is. Do you have an idea of how many colonies you will need?"

Gudrun and Niko are horrified. A chill runs down their spines. Costs for bee colonies were not included in the budget. Professor Ruttner had apparently assumed that the bee colonies would be made available to them. Gudrun quickly looks at the invitation letter to see if there is a corresponding passage on bee colonies. In fact, there is nothing about the bees—only the accommodation is promised. Dung! In no time at all, they estimate what is financially possible. The travel fund allows the purchase of a maximum of four colonies. Hopefully they will be enough for their experiments. And hopefully they won't have to face any more unexpected costs. Mr. Bar Cohen is satisfied and confirms their decision:

"Four colonies. Very good. I will inform the beekeeper today. He will deliver the bee colonies directly to the station in Zrifin. It's hard to pick them up on a motorcycle." He smiles and continues: "Your accommodation is taken care of. We have reserved a bungalow in an orange plantation right next to the bee station. Unfortunately, I don't have time to go there with you. I have to go to the ministry. You understand."

He sighs theatrically. "I asked Rafi, our driver, to take you there."

Gudrun doubts whether this is true with the ministry. In reality, the guy probably has no desire to have anything to do with them, the two Germans. It seems to her that he got the whole thing on the table on order from "above" and had to find a solution with little contact and at a large distance.

Bar Cohen reaches for the big black phone again and says something in Iwrith before turning to them again in English: "Rafi is coming soon. If you will wait outside, please?"

He stands up, opens the door of his office and points unmistakably into the corridor that leads to the barracks exit. They know the way to the parking lot. They don't have to wait. Even before they arrive at the Horex, a small azure blue station wagon followed by a large cloud of dust whizzes into the parking lot. Maybe a little too sporty. The cloud of dust only settles after Rafi, a slender young man about their age, has gotten out of the car and shakes their hands. Niko has seen some such cars on the way and asks:

"What kind of brand is this, Rafi? Such cars don't drive in Germany."

"That's a Sussita! An Israeli car, produced in Haifa!"

They shouldn't argue about design and color. Niko hopes that it will be a reliable vehicle. Gudrun sits in the passenger seat and Niko follows on the Horex. Now things are progressing faster than with Fritz Kuhn. They drive between orange plantations and at some point, Rafi turns and seems to follow the signs in the direction of Jerusalem. One of the first places behind the city is Zrifin, which was called "Sarafand" during the British Mandate. Like Hakirya, Zrifin had also been a military camp of the British colonial army. On the western side of the site, the Israeli army operates a large military complex. Hermetically sealed. High fences, barbed wire, watchtowers. The eastern side of the former military camp has been converted into a large experimental agricultural plantation. A fence full of holes, no barbed wire, no watchtower. That is reassuring. The bee station is also housed here. In the north of the experimental plantation is an important building, probably the largest military hospital.

They turn into the plantation through a large gate and follow an avenue flanked by pecan trees to the main office. Rafi takes the two to the plantation manager, Mr. Hiller. On the way, Rafi explained to Gudrun that Hiller also comes from Germany. After their friendly "Guten Tag " they are greeted by icy silence. Hiller turns away briefly, coughs.

He is visibly trying to regain his composure. After a long pause, he produces a "Shalom". They have not sufficiently prepared all the lessons of the crash course "Israel for Germans" at the Moledet. Later, they will learn that Mr. Hiller is the only survivor of his family, who were killed in Auschwitz. The confrontation with the German language and two Germans touched painful memories. Gudrun and Niko look down at the ground in shame. They will never repeat this mistake. From now on, people they meet for the first time in Israel are always greeted with "Shalom". The German language often awakens terrible, often repressed memories. As a result of these fates of suffering, German has forever disqualified itself as the language of the master race, oppressors, torturers and murderers among many Israelis.

Mr. Hiller catches himself and politely asks in English about their trip, how they managed all this on the motorcycle. Finally, he pulls out a large old key lying on his desk: "This is the key to your house. It is right next door. If you just follow me?"

Outside, they pass a group of plantation workers who look at them curiously. Gudrun waves and everyone is happy. The house in which they will live for the next six months has thick masonry, a flat roof and small windows. In the style of the British colonial troops after the First World War. At first, they are thrilled: a whole house for them alone! What a difference to Niko's small room in Bockenheim. Mr. Hiller hands over the keys and says goodbye in a friendly manner with another shalom, which they now reciprocate beaming with joy. In the building, the enthusiasm quickly cools down. The house has apparently not been inhabited for some time. As biologists, the spider webs do not disturb them, but everything is dusty and dirty. Rafi, who has come in, wipes the table critically with his index finger. The trail leaves a small pile of dust. He looks at the two and wrinkles his nose. There is a table, two chairs, a closet and a wide bed without bedding. In the kitchen they find a refrigerator, a gas stove and a pot. They had hoped for a little more equipment.

"Looks like we'll have to limit our cooking to one-pot in the next few months."

"Plates are sometimes practical when eating," sighs Gudrun.

Rafi brings them the backpack out of the car. He first wants to put it on the table, sees his finger mark there and decides on the bed. Gudrun and Niko unpack the two saddlebags. It gets loud outside. The plantation workers are discussing loudly and excitedly with Rafi. Is it about the German students?

Rafi now takes the two to the bee station. There they meet the director, Mrs. Alpern and the beekeepers Mr. Grozs and Mr. Sushinski, who greet them in a friendly but reserved manner. Mrs. Alpern gives the impression of a well-groomed lady: Oxford English, immaculate white, starched and ironed lab coat. She is the competence in the laboratory. Grozs, like Ms. Alpern, comes from Poland. He is a slim, sinewy man around 50 years old, thinning hair, tanned, deep smoker wrinkles running through his face. The white jumpsuit has many stains. The bee veil is stuck under the arm. He speaks German with them, with a Yiddish tone. Sushinski is a tall, athletically built old man with a white-haired tousled head. As Mrs. Alpern will say later, no one knows how old he really is. Sushinski came to Israel after 1945 without papers and lives with a distant relative in Tel Aviv. He also speaks German enriched with Yiddish expressions. They don't understand everything, but they understand it better than Iwrith. Sushinski immigrated from Brest in what is now Belarus. Before the war, it belonged to Poland. In the bee station, he seems to be the man for the rough, regardless of his age.

Ms. Alpern asks with interest about her research plans. That will have to wait: "We are

53

sorry. Necessary things are missing in our house. We have to go back to Tel Aviv first and buy provisions, kitchen utensils and a blanket for the night!"

They drive towards the high-rise building, past many small shops and street stalls. Because of their communication difficulties and because they are tired, they don't buy anything there. There is a self-service supermarket in the high-rise building, which is easy and convenient. They buy two plates, two cups and two forks, spoons and knives each, always the cheapest offer. Nevertheless, it is expensive and the money is only enough for a thin blanket. Later they will find out that the prices in the small shops at Carmel market are much cheaper. Big shops, small prices do not apply in Tel Aviv.

When they open the door the next morning to look around, Gudrun and Niko find packages and objects of all kinds in front of it: a pan, a kettle, even an armchair and many useful items that they can use in their new household. They run to the bee laboratory and learn that it was the plantation workers who brought them the things. Mr. Grozs states:

"It's not possible that two young people like you don't have enough!"

Gudrun and Niko are touched. As they carry the things into their house, Niko puts it in a nutshell: "In Oberursel, farm workers and beekeepers would hardly have done that, for any doctoral students from abroad. They would have made fun of how primitive they would have to live." Gudrun nods: "Yes, it's different here. Remember what these people experienced and suffered. These are Jewish immigrants. They know what it's like when you arrive with nothing. They didn't even have the saddlebags on their motorcycles like we did when we arrived here. They simply help when they see that help is needed!" "Yes, that's right. The fact that they are helping two Germans, of all people, is something special!" Gudrun is silent. Niko too. History catches up with them everywhere here. The shortage in the budget has been remedied.

Tenth chapter

Gudrun and Niko begin beekeeping and hear about the Displaced Persons after the war

After they are finished in their new house, they go back to the bee laboratory. There, Mrs. Alpern first explains the situation of beekeeping in Israel:

"When the State of Israel was founded, there were almost only bee colonies of the native bee breed *Apis mellifera syriaca*. These very dark bees are not suitable for commercial honey production and pollination. They are simply too aggressive to be used on the large orange plantations. The plantation workers have to wear protective clothing at all times—that's not possible in the heat. We have therefore imported gentle Italian bees (*Apis mellifera ligustica*) from the USA, and no one is stung."

"The bee colonies that Mr. Bar-Cohen wants to get for us will hopefully also be Italian bees?" "Yes! I know the beekeeper personally. At the beginning of next week, he will bring the four colonies over. Where should they be placed? Behind your house in front of the fence would be a good place. Far enough away from our apiary here. You can prepare four stands there. Shall Mr. Sushinski help you?" "That's not necessary, we like to set them up ourselves. Couldn't we help out at your apiary in the next few days until our own test colonies arrive?" They are very curious to get to know Israeli bee colonies. Are they managed differently here than in Oberursel? Mrs. Alpern seems surprised at their eagerness to work and happily agrees:

"Before that, I would like to know more about your plans. The ministry forwarded your application to me. Gudrun wants to find out how the queen can successfully store the sperm she has received from the drones in her spermatheca for years. Almost every queen breeder has wondered how the queens can keep the sperm alive for several years! Why hasn't anyone investigated it yet? And Niko wants to know how the queens distinguish between drone and worker cells when laying eggs. I've never thought about that before, so I'm particularly interested in it."

She pauses for a moment: "How do you find out? What are your experiments for this?" Gudrun answers first: "That's easy, we have to produce a lot of queens, and I collect the fluid of the spermatheca in a glass capillary and freeze it. Last summer I froze the queens and all the spermathecae burst when they thawed –there was nothing left for an analysis of the fluid! I was very desperate. Luckily I have a new chance here!" "Do you know how to produce queens? Maybe we can help?"

Niko laughs: "In Oberursel we helped with queen production and maybe things are done differently here. If we may, we'll watch you first. For my experiments, I want to observe the queen again to see if she also places her front legs in the cell before she turns around and lays an egg inside. I want to record what she does when she only has drone combs. Then comes the hard part, at least I don't know how to do it yet. I want to prevent the queen from inspecting the cell and see if she can still distinguish between the large and small cells." "Well then, good luck!" wishes Mrs. Alpern. "We are happy to help you if you have difficulties. See you tomorrow with the colonies. Pan Grozs is looking forward to it."

The next morning, they start early. The inspection of all fifty bee colonies is pending. All queens have to be checked to see if they are still the "original Italians" from the USA. Working with the free-standing "magazine hives", which are manipulated from above, is easy. However, with multi-level magazine hives, you first have to lift the heavy combs with honey. It sometimes jerks. Then the bees try to defend their colony. However, working on the lower super with the brood nest from above is really fun. The yellow bees from the USA are indeed very peaceful. Gudrun and Niko are thrilled: If their four colonies are like these, the experiments will work without any problems.

Gudrun helps Mrs. Alpern in the laboratory and with queen production. Niko joins Mr. Grozs. Together they examine the bee colonies, remove and add new combs or feed the splits with sugar syrup. "Four eyes see more than two! The search for queens is really faster in pairs," Mr. Grozs is happy with Niko's help.

Mr. Grozs is skilled. He seems to have two right hands. Niko is surprised how quickly the inspections of colonies are done. And no one complains that people smoke too early or too late. In fact, the smoker is rarely used as Mr. Grozs is a chain smoker and produces enough smoke to keep the gentle bees in check. The cigarettes stay lit during the small breaks: "I don't smoke more than three packets a day!"

That's no problem for Niko. They usually work outdoors and the wind from the sea blows away the acrid smoke of the cheap "Three Asses" cigarettes. Apart from that, Niko believes that three packages a day are not enough for Mr. Grozs.

Cleaning and tidying up in the tool shed is also on the agenda.

Mr. Grozs wants to hear how and why Gudrun and Niko live in Frankfurt and why they came to Israel. During the many cigarette breaks, there are lively conversations as questions and answers line up one after the other. "Nature created bees, and they do what they want. Isn't such knowledge enough to work with them?"

Niko goes to great lengths and tries in vain to convince him otherwise. Private matters are also discussed. Niko tells their story, of meeting in Freiburg, working with bees in Oberursel and the rushed marriage. The story of their honeymoon in the town hall tower prompts Grosz to also report on his eventful life: "The travel destinations in Germany seem to be more manageable and safer these days than twenty years ago." He laughs briefly and dryly: "Do you know what DPs are? This stood for "Displaced Persons." After the end of the war, there were millions of people in Germany who came from all parts of Europe. Prisoners of war, forced laborers, Jews and gypsies from the extermination camps. The Allies did not know what to do with them. So, they left them in the Nazi camps for the time being. But the liberation had come too late for many. They were too weak, too sick, which is why many died in the camps under the Allies." Niko hangs on Mr. Groz's every word.

"After the liberation from the camp in Dachau by the US Army, I was suddenly no longer a Jew as I had been with the Nazis. Now I was a DP. Even though I had escaped certain death as a result, I quickly realized that freedom looked different. DPs were sorted by the Allies by nationality and not by denomination. I was taken to a camp in Poland, although I objected."

Mr. Grozs falters in his narrative, struggles with his composure. He stubs out his third cigarette: "We still have to inspect to the last twelve nucleus colonies!"

Niko can't find the right words to express his sympathy. He understands that a continuation of the description is not possible now. Two days pass before he hears the continuation of Grozs' story. To Niko's question about how the imprisonment of the German army and the liberation by the Americans in the spring of 1945 were experienced by the affected camp inmates, there was a sobering answer:

"We were outraged. Even our smallest hopes were disappointed. We were still crammed into a camp. Instead of the SS, we were now guarded by mercenaries of the Americans. As before, the food distribution was carried out by Germans who worked for the Americans. The rations were better, still not sufficient. Many in the camp were sick. There were no doctors and some of us were dying every day!"

Mr. Grozs lights the next cigarette. Will he continue to speak?

"Above all, there were no shoes. My slippers had worn out. It was very cold. This is not how we had imagined the liberation!" Even after so many years, the deeply felt indignation shows on his furrowed, wrinkled face. He closes his eyes for a brief moment. He looks at Niko in silence for a few minutes. What thoughts are going through his head? Does he wonder why this German wants to know what happened back then?

Then Gudrun comes and asks if they have finished the list with the records of the nucleus colonies Mrs. Alpern sent her? Mr. Grozs swallows, smiles, and then with the polite compliment that is probably customary in Poland: "We'll do everything for the beautiful ladies!" He takes the list out of his folder and hands it to Gudrun with an implied bow. "Now Niko, break is over!"

He stubs out his cigarette butt on a stone on the floor, and they go into the shed where they continue to cut the old black combs out of the frames. The frames must be scraped clean. A monotonous work. "Were you in the camp for displaced persons for a long time? How did you come to Israel? Was that officially possible with the help of the Allies?" "Oh, Niko. That's too many questions at once. Certainly not with help, but against the Allies. In this camp we were no more than 50 Polish Jews and certainly more than 500 Poles, Christians. We Jews wanted to go to Palestine, not Poland. The majority of Poles also did not want to return to their Communist homeland, which was now occupied by the Russians. No one there was stupid. Of course, we Jews in the camp were quickly organized. The Poles and other groups also worked with us. Everyone had a common goal. Out of this camp and out of Germany." "Did you break out?" "Yes, luckily it was unspectacular and well organized. This was arranged by a group of our Haganah. The guards were probably drunk or bribed. In any case, three trucks came into the camp at night. Only for Jews. We jumped on and off we went. I arrived with eight comrades to village in Alsace. There we got clothes, shoes and enough to eat. We helped with the harvest and there I got to learn about bees for the first time. That helped me get back to life." Mr. Grozs lights another cigarette.

"How did that happen?" Niko wants to know. "Niko, I'm sure you know that yourself. Bees fly all day long. They start empty and come back heavily loaded. Do they know why they're doing it. They just do it. Bees certainly enjoy working for their colony." Niko understands: Often it just comes down to doing something good! The why is initially less important. What can he answer Grozs? His own life in orderly, even sheltered ways in post-war Germany is too far removed from the sufferings and experiences of Mr. Grozs. All he can say is that he also feels the attractiveness of the bees. Will that be enough? "Niko, you are silent. You are right. One should not be tested! You cannot know how we Jews fared at the end of the war, shortly after we had survived the concentration camp."

An empty cigarette pack is crumpled in his fist and tossed purposefully into the trash can. With shaky fingers, the next package is torn open: "Niko, go ahead. I'll be right back!" With a burning cigarette in the corner of his mouth, Pan Grozs goes out to the apiary. He crouches down next to the entrance and watches the bees as they fly in and out with eagerness.

The next day, Mrs. Alpern asks Niko and Gudrun to help her:

"We need new mating nuclei for the queens that emerged yesterday."

"Go ahead, Niko. You know how to do it," Mr. Grozs tells him. "I'm going to the comb storage shed." Did he ask Alpern to take the curious Niko off from him? Did his questions hurt him? There is no time for thoughts. Niko has to hurry. Gudrun needs the nuclei. He quickly inserts two brood combs with bees, another with honey and a fourth with pollen into a brood box and was done.

No, Grozs is not upset. On the contrary: shortly before the lunch break, he beckons Gudrun and Niko to him. He brought them a large avocado from home:

"It's almost pure fat, lots of vitamins and very healthy. Just the right thing for a beautiful young couple like you!"

Gudrun smiles back happily and thanks him for the beautiful fruit. She doesn't tell him about the avocado salad at Pardess Hanna, which she didn't like. "With your compliment, it will taste particularly good to us!"

Mr. Grozs then becomes more serious. "Gudrun, your husband told me that you grew up in Wilhelmshaven and also that you had an exciting honeymoon to the town hall tower. I've stood on it and looked out to the sea". "Unbelievable! How did you get there?" "After my escape from the Dachau camp, I didn't stay very long in Alsace with the bees. The Haganah had organized a crossing for me from Marseille to Palestine. I received travel documents and tickets that were so well forged that the train ride to Marseille went without any problems despite multiple checks by customs and police.

Then came the embarkation. The ship was named "President Warfield". I still remember it like today: it was July 10, 1947. We were over 4500 passengers on board. Many children, most of them orphans. It was terribly cramped on the small ship. It departed at three o'clock in the morning. As soon as we were at sea, we were accompanied by three British warships. The British had apparently gotten wind of our destination. As long as we were in international waters, however, they were not allowed to board us. This went on for seven days across the Mediterranean. On July 17, off the coast of Palestine, our ship was ceremoniously renamed "Exodus 47" and the Israeli flag with the Star of David was hoisted. Everyone cheered."

Sadly, he continues: "We knew that we had no chance of ending up in Palestine unnoticed. The next day our ship was rammed and captured by the British. There were even deaths. We were towed to Haifa and taken directly to three British deportation ships. Nobody knew where it was going, only one thing was certain, it was not Palestine. It was terrible not to be able to go ashore after all the deprivations. It was to get worse: We were taken back to Germany. To Germany! The land of murderers. We were interned there in the same Nazi camps where we had only escaped death by good fortune. Again, we were transported on barred railway wagons, again housed behind barbed wire. I ended up in an old SS camp near Lübeck. There was no heating and I was transferred to a camp near Wilhelmshaven. I was there for several months. We had enough to eat and were even allowed to leave the camp. We just had no papers and had nowhere to go. The Haganah did not remain idle and organized new papers and new transport plans to Palestine for me. From there, the whole thing started all over again. This time via Holland to Marseille. Again, on a boat with a thousand others. This time the crossing succeeded without a British ship escort and we successfully landed on the coast of Palestine. We had to wade through the water for the last few meters. It was wonderful to lie on Ashdod beach in the middle of the night, completely soaked and exhausted. There I was born again. Shortly thereafter, in May, the State of Israel was founded and the British withdrew. I became Israeli and was no longer a Pole."

Mr. Grozs lights a cigarette, takes a deep drag: "And why am I telling this story now, which should never have been told again? Yes, the bees are to blame. It was the bees that brought both of you here!" He laughs at the end, that sounds bitter. Gudrun is shocked. She was then 5 years old and nobody ever told her about the camp in Wilhelmshaven, her town!

During the weekend, the bee station is closed. The conversations of Gudrun and Niko revolve around the unimaginably terrible conditions that continued in the years after the German surrender. The conversion of the concentration camps into camps for DPs. Instead of SS guards, there were now Allied guards. Still no freedom. Still on food rations. Mr. Grozs was certainly not an isolated case. Much his story overlapped with that of Fritz Kuhn. Hundreds of thousands and more shared the same terrible fate. Was this unbelievable approach by the British only due to the chaos in the first phase of the occupation? Or was it the continuation of European anti-Semitism by other means? In contrast to the Holocaust, which was discussed in detail in the curriculum of schools in the FRG, only now they have learned about the DPs and their fate at a bee station in Israel. Gudrun and Niko are shocked that these tragedies were concealed. They get angry when they remember how the questions to their parents were always dismissed with "there was war" or worse, with "that was so long ago." Here in Israel, everything is present. How much more dirt is still under the carpet

Eleventh chapter

Gudrun and Niko don't want to be Dutch and meet Meir

Life away from bees remains a challenge, even if they do not report it to Germany. In Germany, they exchanged their travel funds for traveler's cheques, issued in US dollars. This means that they are largely protected against theft and loss. This security comes at a price. Traveler cheques can only be exchanged in a bank into Israeli lira for a fees. In addition, they must wait a long time at the bank in Rishon until all the formalities for this unusual procedure have been clarified. Over time, the employee at the bank counter for foreigners, a young man their age, knows them and they talk. He asks Niko if he is a Christian? This question, Niko thinks, goes too far and is too personal. He has since learned that in such cases a counter-question can be the best response. "Do I look like a Christian?" he asks back in a loud voice. The reaction causes him to turn red: "Oh, no, sorry very much. You don't look like a goy at all. Your face, your eyes and overall, no, you look very Jewish!" Niko laughs and rejoices with the bank employee about his Jewish appearance.

Due to the purchase of the bee colonies and the additional necessary household utensils, their budget has shrunk. They have less than 10 dollars per week. They have to shop and the supermarkets in Tel Aviv are expensive. It is cheaper in nearby Ramla. The Kuhns warned them about Ramla. There you can get seriousfood-borne diseases. The budget regulates their shopping. At first, they cannot get used to the Arabic-influenced way of selling meat. The goods hang outside in the sun and are densely covered with flies. In the first few weeks, they live like vegetarian, from falafel with salad in flatbread. But after some time, they miss animal protein. Finally, their lust for meat wins out over the tight budget and they drive to Rishon le Zion, where mainly Jews from Europe live. There, the butcher has a refrigerated counter and air conditioning, and the flies stay away. In a meat frenzy, they buy a whole chicken. This is cheaper than anything else and is enough for a whole week. On the last day there is a soup made from boiled chicken feet. Over time they become familiar with the oriental market, and buy a weekly chicken in Ramla.The narrowness of the market and the crowds are simply part of their experience.

The cultural difference between the "Western" Rishon le Zion and the "Oriental" Ramla is unmistakable. It only takes a few kilometers to travel from the Occident to the Orient. Without any fence or border, the cultures separate. Ramla has a memorable history in the founding of Israel. From there, Palestinians frequently undertook raids and attacks against Jewish settlers before and during the military conflicts in 1948 when the State of Israel was founded. After the victory of the Israeli armed forces and the extensive expulsion of the Palestinian population, the majority of Jews from the Middle East and North Africa settled there, some of whom were expelled from Islamic states. This has created a colorful, very diverse community that is becoming more and more attractive to Gudrun and Niko not only because of the low food prices. Depending on the origin of the merchants, they speak French, English, but only rarely German. They also learn some Iwrith. The business people in Ramla are consistently friendly and helpful. The two develop a special friendship with Bela, the pharmacist. He is an old, small, white-haired man with flashing eyes under bushy eyebrows. When he hears that the two are working on the biology of honey bees, he spontaneously takes them to his heart. He tells them about his pharmacy studies in Budapest, and how he too fled from the persecution of the Jews, but only after graduating as a pharmacist.

"Actually, I had wanted to do a doctorate, but history was against it. It's good here at Ramla. Here, origin doesn't matter!" He follows the stories about their research projects with great interest. Yes, back then in Hungary, they also had bee colonies. My father took care of them and he was allowed to help. Then pharmacy seemed to be the more rewarding subject to study.

They soon feel at home in the plantation. Their relationship with Mr. Hiller, the plantation manager, gradually eases. He is probably relieved that there are no difficulties between the German students and his workers. Yes, probably to his surprise, even a number of plantation workers, a colorful mixture of European and North African Jews, have made friendly contact with them. Greetings and short conversations, often in Yiddish, are part of the everyday life of the young researchers. Finally, Hiller also begins to take an interest in their experiments. He drops by from time to time and talks to them, now even in German. He explains to them his experimental program on the plantation and the tests with the new orange varieties. During one of these conversations, he invites the two of them to come to his home with his group of friends.

Gudrun and Niko are happily surprised. Has the ice been broken? He asks: "I can't introduce you there as Germans. You have to understand that. I will say that you are Dutch." It becomes clear that many of his friends cannot meet and talk to Germans. Cautiously, Gudrun explains that they understand well that there are many people in Israel who, after everything that has been done to them, want nothing to do with Germans.

"It wouldn't be good to pretend to be Dutch. This is risky. All it takes is one wrong question and everything is exposed! Then we are not only Germans, but also fraudsters." Mr. Hiller looks at Gudrun in surprise. He hadn't thought of that. Only one specific question about the occupation of the Netherlands or the extermination of the Jews and Gudrun and Niko would be exposed. He nods and remains silent thoughtfully. After a long pause: "You're right, that's not possible. I'm sorry I didn't think this through to the end. Large parts of society here in Israel are not yet ready for such an experiment. Nor was I myself before you came to Zrifin. Over time, however, I have come to know and appreciate you. I am extremely sorry that not everyone here is yet ready to recognize that the youth of the Federal Republic of Germany has successfully cut the cord from the Third Reich in such a short time." His eyes become moist. The dark history has caught up with him again.

In the evening it gets lonely on the experimental plantation. Only Meir, the night watchman, arrives there every evening at 7 p.m. sharp and stays until seven in the morning. He is a small, slender man, around 60 years old, and he regularly visits Gudrun and Niko. Meir was expelled from Iraq in 1951 with his wife and seven children and came to Israel with "Operation Ezra and Nehemiah." Since then, he has lived in Ramla. His eldest daughter married in Israel. Meir proudly talks about his eleven grandchildren. Because his younger six children successfully completed schools and their vocational training, mostly during military service, they have good incomes. To his regret, not all of them have married.

Meir comes from Basra, according to his reports, the most beautiful "pearl" of Iraq, the most beautiful city in the world. He would have liked to stay there, but after the founding of Israel, it became difficult for Jews in Iraq. There were repeated bomb attacks on Jews. Initially, there was a ban on emigration. That wasn't bad, why should you leave Basra? Then it became clear that Iraq wanted to get rid of all Jews. They were now allowed to leave the country, but they had to give up their Iraqi nationality. As more and more bombings followed, Meir also had to deal with fear. He registered his family for emigration and became stateless. Iraq deprived them not only of their nationality, also of all their property. It was a cruel time. In the end, the Israeli government regulated their emigration. A flight from Basra to Tehran, and from there to Tel Aviv. Now he had to start all over again. Ramla is of course not quite as beautiful as Basra, but now he feels comfortable there. It's good for the family, and what's good for the family, is good for him. It's good in Ramla. It's that simple. It's also not as expensive as in Tel Aviv or Rishon le Zion. "We have noticed that and have always shopped in Ramla in the last few days. In addition, everyone is very friendly and helpful when we don't know exactly what's good."

Meir beams: "Yes, we try to help each other. That's important in Israel!"

Whenever their time allows and no reports or letters have to be written, Gudrun and Niko sit together with Meir. Gudrun usually invites him, her kitchen and also the large table now are reasonably clean. He often brings delicacies with him, prepared by his wife: chicken necks stuffed with rice, spicy and seared, pilaf with lots of vegetables and spices and a little mutton. After dinner, they follow his lively stories from the Thousand and One Nights with the greatest pleasure. His language is flowery and lush. He indulges in superlatives.

He insists that they tell him German fairy tales. He is impressed by the Grimm fairy tales they tell. Instead of rich royal houses, the story usually talks about the poor, who succeed and become rich through a good heart, coupled with courage. Meir enjoys the fact that he is no longer alone on the ward all night. After meals together, he often talks about invitations to the best restaurants. He raves about sumptuous oriental feasts, which he describes in every detail. At first, Gudrun reacts very affected: "Niko, that's not possible. We know that Meir is poor, and now he wants to spend his last lira on us."

Niko shakes his head: "Hmm, I don't think he means it that way. It's like his fairy tales from The Thousand and One Nights. More an expression of Meir's appreciation for us. This must not be misunderstood as a promise or obligation. A polite sign to express a good mood appropriately."

"Maybe we should engage a little more so as not to be rude? I'm afraid that can only sound wrong!"

In any case, Niko was right, these invitations are never concrete. But one evening it is different. Meir is excited: "My youngest daughter wants to get engaged. I cordially invite you to this celebration."

His face beams: "The day after tomorrow I'll pick you up in a taxi at three o'clock in the afternoon. Just wait for me on the road from Tel Aviv to Ramla. I have to be back on the ward at seven in the evening. We can return together."

They enthusiastically accept this invitation. "We are happy to come. We look forward to meeting your family."

What do you have to do at an oriental engagement party? They don't want to make mistakes. The next morning, they ask their friends in the station what kind of gift would be appropriate for an engagement. The suggestions are broad and range from jewelry to chocolate. Someone suggests perfume, and then everyone thinks it's kind of good. So, perfume. They drive to their pharmacist friend Bela in Ramla and make their request to him. Bela is delighted: "Yes, perfume is a good gift. My tinctures smell good too. Actually, only a few. However, in the shop opposite Shafi there is an even better selection. I'll show you."

Bela locks his pharmacy and hangs a large "I'll be right back" sign on the door. Together they weave their way through the heavy traffic. Shafi's shop is small but richly decorated with large-format black-and-white photos of oriental beauties. In the shop, the beguiling scents of the Orient spread in an unimagined concentration and overwhelm Niko—he is surprised that the limited oxygen pressure still allows breathing. Shopping has to be done quickly. Gudrun decides, and Bela agrees, as Niko is too infatuated to be able to contribute to the decision-making process in a meaningful way. After a short negotiation, the price is also right. The chic bottle is extensively packaged and decorated with a golden bow. Niko has left the shop to catch his breath. "Well done! Have fun at the party!" Bela calls after them as the Horex drives back to the station. The next day it gets serious. Gudrun puts on her only skirt and chic blouse: "They were intended for ministerial reception. I hope it's enough for the engagement." "The girls in the photos at Shafi's had a different outfit. What should I do? I only have a white shirt. At least something. The scuffed jeans are certainly below any standard."

Then both are standing on the road. At three o'clock sharp, a taxi stops at the entrance to the station. Meir sits beaming in the passenger seat and waves excitedly at the two. They get in the back and drive through the crowd on boldly winding alleys to Meir's apartment in the center of Ramla. in the entrance to the house they hear loud laughter and also singing in the high notes typical of Arabic music. This completely different way of singing is no longer foreign to them. Meir always plays his small transistor radio at full volume during his nightly tours of the station. Oum Kalthoum is his favorite singer. In the two small rooms about 50 good-humored guests are together. Meir introduces Gudrun and Niko to his family. They congratulate the fiancés and wish "Ad mea esserim" until 120! Everyone laughs. Laughter, tears! Another mistake? At least not everyone has fallen into icy silence and turns away, as usual when things go wrong. They quickly learn that this is how you congratulate someone on their birthday, never on your engagement. They laugh along and put their package with the gift on a small table, where many gifts, mostly household appliances, are piled up. Then some guests slide together and they squeeze in between. They get small plates of very sweet cakes and a cup of sweetened tea. They are inundated with questions: "How do you live as a student in Germany? How much does a Volkswagen cost there? Are all the motorcycles there as big as yours?"

They try to answer everything and always earn a lot of applause and laughter. Finally, Meir turns the conversation to Niko. "A handsome man, so slender and yet powerfully built. Such beautiful blue eyes! Strong shoulders." Niko's attempts not to blush fail, as do his efforts to end this embarrassing discussion. It is only after half an hour that Meir notices that Gudrun has become very quiet. Now a compliment for Gudrun has to come: "Niko may be very beautiful, but Gudrun, she has got a good heart." Everyone applauds, the company goes wild. From the radio comes another song by the famous Oum Kalthoum, which everyone sings along loudly and with full commitment. Then bowls of new food are served. The food is now more important than the conversation about Niko's beauty and Gudrun's good heart.

In Zrifin, Gudrun and Niko are far removed from the public and media events in Israel. Being so isolated is not perceived as a restriction by either of them at first. Too many new impressions and oppressive stories from the German and Jewish past assail them. In addition, they can concentrate fully on their research work in this environment. Some time passes before Gudrun and Niko try to gain access to the current political environment. Meir initially plays a decisive role in this. He reports, often in very drastic terms, of terrorist attacks and the shelling of peaceful Israeli settlements in the north of the country by Palestinians. For Meir, the Palestinians are "evil terrorists." He is happy about every successful retaliatory strike by the Israeli soldiers and announces martial slogans:

"All Palestinians should be driven into the sea and destroyed!"

It is clear to Gudrun and Niko that Meir, when he speaks in this way, is repeating statements made by Arafat and other radical leaders of the Palestinians. With all due respect–Meir as the only source of information on the Palestinian conflict, that really doesn't work. The small transistor radio doesn't help much. Iwrith or Arabic. On one of their shopping trips in Ramla, they ask for a newspaper. Yes, there are various Israeli newspapers in non-Hebrew script. The "Jerusalem Post" is English and unaffordable for Gudrun and Niko.

Twelfth chapter

Gudrun and Niko are accused of espionage

On the next visit to Pardess Hanna, Gudrun and Niko tell Dwora and Fritz/ about their news problem:

"Everything we learn about day-to-day politics in Israel, Meir the night watchman tells us. And these are mostly bloody horror stories, presented in a very one-sided way."

Dwora has the solution and digs out a stack of the German-Israeli newspaper "Yedioth Hadashot", to which they have subscribed.

"That's a good thing. I actually wanted to throw them away, but it's much better this way. Of course, I will be happy to keep the newspapers for you. You can then take them with you on your next visit."

"We are very pleased about that. We also want to learn about the daily events and political background in Israel."

Dwora gets a few strings and packs a bundled newspaper package. Over coffee in the afternoon, Fritz talks about his adventurous tours through Galilee, when he has to check his sales outlets for drinks. Exciting stories as always, not as flowery as Meir, but sober, clear and almost ready for print. Gudrun is impressed.

"When are you going back to Galilee? Niko and I would like to come along if you have room in the car. We have a short break now anyway while we wait for our queens to hatch."

Fritz happily agrees: "That's a great idea, and I know what I'm going to show you. Tomorrow is Sunday, a Christian holiday, as you know. My VW workshop in Nazareth belongs to Abu Hassan, a Christian Palestinian, and he then closes the shop. The Beetle needs an inspection, so I have planned my trip for Monday. Just stay here until tomorrow. A day at the Mediterranean in Cesarea is not bad either. Then we'll go north together on Monday." Fritz's suggestion sounds seductive and like adventure. Niko looks at Gudrun and dashes briskly forward: "That's a great proposal. Can we abuse your hospitality for so long? For us, the one day does not matter. And sea plus Galilee is better than Galilee without sea?"

Gudrun agrees: "To the sea is great. I always have my swimwear with me. Whether it's the Sea of Galilee or the Mediterranean: swimming trunks and bikini are good for both!" Dwora and Fritz are happy that the two are staying with them. After planting the wild lemon "Chrush Rash", they talked until midnight. After a rather late breakfast, Fritz gives them a long lecture about the cold of the Mediterranean, the murderous currents and the particular danger of the coast near Cesarea. Then the phone rings. As far as they can understand, Fritz has forgotten an appointment at the lemonade factory. A kiss for Dwora, to his beetle and off to the meeting. Gudrun and Niko take a little more time. Dwora insists on packing them a thick lunch. They swing onto the Horex and are on the coast just a few minutes later. In the middle of the dune landscape, they are suddenly surrounded by Roman and medieval walls everywhere. The road leads through the middle of the dunes, unthinkable on the North Sea coast. Passing the ruins of old fortifications, they arrive at an aqueduct from Roman times. There are still hundreds of meters along the beach. A plaque informs them that the aqueduct supplies the nearby port city of Cesarea with water from the Carmel Mountains, six kilometers away. There they are now in the middle of archaeology. Gudrun's father has pestered her enough in Wilhelmshaven: "Niko, let's get into the water. The old stones can wait even longer."

The sea is warmer than the North Sea in summer. This seems not warm enough for everyone. They are the only swimmers. A lonely beach hiker passes by and explains to them with a worried expression that water temperatures below 20 degrees are harmful to health and very dangerous. Obviously, he never swam in the North Sea.

They also went to the 2000-year-old port of Cesarea. Gudrun finds Phoenician shards of glass in the sand: "Just look at it, Niko, they were also found in my father's excavations along the coast in Germany. Around the birth of Christ, trade reached us on the North Sea." Niko answers succinctly: "I didn't find them in Bad Salzuflen." He is more impressed by the honey bees that collect at the garbage can:

"Look, in Germany only wasps do that!"

Next Monday, they will start early in the morning in Fritz's Beetle. They pass Mount Tabor. Fritz asks whether he should show them the monastery on the mountain or continue without stopping for a morning coffee break in Tiberias on the Sea of Galilee. Gudrun, who sits in the front next to Fritz, has a clear opinion:

"Better a freshly brewed coffee on an old lake than an old monastery on a fresh mountain!"

Thus, they drive on in a good mood. Then Niko asks about the security situation and reminds them of the horror stories they heard from Meir:

"Is this oriental exaggeration?"

Fritz frowns. "Unfortunately, not always. Especially on the northern border, there are still huge refugee camps with Palestinians in Lebanon, very close to the Israeli border. They have no work in the camps and have been staying there for almost 20 years, supported by United Nations funds! There is no future there, because an entire generation has grown up in the meantime that knows nothing but their status as refugees. Refugee by profession. That doesn't make anyone happy. The Fattah Palestinian political party has an easy time spreading hatred and violence. It is so easy to blame Israel for the poor living conditions. The British, who caused the expulsion with their colonial policy, are far away and no longer there. They did not want to integrate the neighboring Arab states, which even expelled the Palestinians from their villages in the 1948 War of Independence. They lost the war, but did not take in the displaced persons. After that, many Jews were expelled from the neighboring Arab countries, of whom many were taken in by Israel. Just like your friend Meir. In the meantime, Israel has been recognized as a state under international law. The Palestinians are not interested in that. They are used as bargaining chips to make political profit. While the expulsions in Europe after the war did not destabilize the political situation, this has failed thoroughly in Palestine. Here, the complete annihilation of the State of Israel and the reconquest of the land is a priority goal for most Palestinians. That was the case before the war. Arafat and his terrorist organizations will not rest, but continue inciting everyone. The UN soldiers stationed on our northern border offer no protection. They are ignored by the Palestinians. And when the situation becomes critical, they run away! Israel is dependent on its own armed forces. Israel's short history has shown that only their own soldiers can protect the country."

They reach Tiberias and the Sea of Galilee. Fritz stops in front of a small café with a terrace overlooking the lake. They sit down at one of the tables, just a few meters from the water. The owner, an older man, very corpulent and with a sweeping moustache, greets Fritz very warmly. They seem to know each other and disappear into the back of the kitchen. Apparently, Fritz is planning something. He comes back to the table, together with a young waiter, and brings them two large plates of something unknown, coffee, and a glass of water.

"I've known Ben for a long time. He makes the best fatteh in Galilee. Bread baked in fat, with hummus and various spicy sauces. Ben's sauces are simply outstanding! Fatteh is a traditional oriental breakfast. Terribly fat and even more unhealthy! It tastes good, ohoh!" He raises his hands enthusiastically.

"However: the cholesterol, it's too high, so because of my cholesterol I'm forbidden fatteh. Unfortunately, the only thing left for me is coffee."

The fatteh tastes excellent to the bee researchers and their plates empty faster than their coffee cups. Fed up, the journey continues.

"Our next destination is Kibbutz En Gev on the other side of the Sea of Galilee. En Gev often has problems. The Syrian army has shelters on the Golan Heights, where Katyusha rockets are fired again and again. Artillery and tanks are also stationed in the Golan. Their bombardment is much worse. Fortunately, it has been a bit quieter lately. I have an old good friend there, Avram Selikowitsch. We arrived in Israel together. He is one of the founding Chawerim and will show us how to live and work in the face of such a threat."

They cross the Jordan River, which is fed by the Sea of Galilee. Gudrun is disappointed and cannot believe that this pitiful trickle is supposed to be the Jordan River, which she has heard about in the children's service:

"I had imagined crossing the Jordan River to be more dramatic," she grumbles.

Fritz laughs at her disappointment. "Yes, your Rhine is much bigger and much more beautiful! Here in the desert, every drop of water is priceless!"

They continue along the eastern shore of the Sea of Galilee.

"Here we are very close to the border. There on the right the hills are in Syria. A few kilometers further north, the border extends to the lakeshore. You can therefore only drive to En Gev from the south."

Before entering the kibbutz, they encounter a roadblock. Uniformed men, combat boots, barrett firearms, sunglasses and rapid-fire rifles. Tough guys. They wave the beetle to stop. One runs around the car. Fritz rolls down the window and mentions the name Selikowitsch. This seems to have more effect than identity documents. Their faces brighten. One even takes off his sunglasses. Everything is in the green zone. Friends of

Avram Selikowitsch are apparently welcome.

"Ah, that's like opening sesame," Gudrun jokes.

Fritz raises his eyebrows. He knows his way around well and stops in front of a small bungalow after several left-right combinations. They knock. Nothing happens. Only after Fritz's sonorous cry "Chawer Selikowitsch!" does the door open. In front of them stands a rather neglected old man who greets Fritz warmly. A half-hug that Fritz can't fend off. Then Gudrun and Niko are introduced and immediately surrounded by an intense scent of alcohol. When Fritz asks him to show them the shelters, Selikowitsch's face brightens. He takes a large key and shuffles in front of them. After a few meters, they stand in front of a large metal door that leads into the hall in front of them. Selikowitsch gives Niko the key. When he takes it and puts it in the lock, it moves smoothly. The castle is obviously kept in good condition. The solid steel door opens easily. Behind it, a short concrete tunnel opens up, reminding Niko of the concrete bunkers of the Bundeswehr. The door closes behind them. With the light of the sparse incandescent lamps, they enter a first room through another metal door. Children's beds and bunks are lined up on the walls, three on top of each other. Further back, they discover a large water tank and packages of food. In an adjoining room, rifles, submachine guns and ammunition are stored locked on large shelves. Fritz asks Avram when they were last fired at.

Avram whines loudly: "That was only yesterday. Bad: Two children could not to be found. Only after the impact of the first two missiles could they be brought to safety here." Avram begins to complain loudly, even cry: "This goes on and on, just like it used to be. You never know whether tomorrow or the day after tomorrow or in a week. It never ends."

Fritz hugs Selikowitsch and reassures him: "Avram, that's different today! We belong here, this is our country. We are at home here and we will win!"

Fritz does not remain calm either. Sobbing, the two men lie in each other's arms.

Suddenly, the mood of mourning is abruptly interrupted. The door is abruptly pushed open. Two young men in uniform are standing in the doorway. Just as martial as those at the roadblock, only without sunglasses. They rant loudly and vehemently. Almost panicked. Niko and Gudrun don't understand anything, Fritz answers in Iwrith just as excitedly. A brief argument ensues. Fritz explains to Niko and Gudrun that they have to go to the main office with the two guards. He would follow with Mr. Selikowitsch. Anxious and distraught, they climb into the back seats of an open military jeep and drive to the main building. The leader of the two men is probably a non-commissioned officer or sergeant, in any case he has more stripes on his sleeve than the other. He takes them to his office and sits down behind the desk, while the other one stands next to the door. Is it important to prevent an escape attempt? The multi-stripe carrier pulls out a form and asks Gudrun and Niko for their names in Iwrith. At least that's what they suspect, because they don't understand him and explain in English that they don't know Iwrith and that he should ask his questions in German or English.

"Names, please! " They put their names on record. Gudrun's resistance returns. She thinks that young men, whether uniformed or not, should behave more politely:

"And what is your name, please? " The two uniformed men are surprised and do not know immediately whether this is a military secret. Finally, the multi-stripe carrier explains himself in English: "My name is Boris. You surely know that you have committed a criminal offense? Spying on defenses. When you entered the country, you signed a form to have taken note of the fact that foreigners are prohibited from entering and exploring military installations. Your IDs!"

Gudrun rummages slowly in her handbag. Boris becomes impatient and asks if they don't have passports and how they got into the country. Niko tells him about the invitation from the Ministry of Agriculture and about the VIP clearance in customs with the help of Mr. Rabinski. In the meantime, Gudrun has found the passports and pushes them across the table. Boris checks page by page very carefully. Obviously, he begins to doubt: No spies after all?

The door opens and a young woman in a chic uniform enters. No stripes, but stars on the collar. In her early 30s, apparently with a higher rank, used to listening and being heard. Boris and his colleague stand at attention as they greet the woman.

"Welcome to En Gev. My name is Gila. I'm responsible for security here. We had a bad security breach yesterday. Despite early warning, not all children reached the shelter in time. Avram Selikowitsch has told you that. And now you burst in here and take a look at our shelter. This is strictly forbidden for foreigners. And you are Germans," she waves their passports, "and therefore foreigners. That's bad. I would now have to detain you in En Gev until a possible espionage background is ruled out. I want to spare you that. I spoke at length with Fritz and Avram and learned that you work in Zrifin. We can reach you there at any time if we have any further questions. However, I ask you not to leave the country within the next week."

Gudrun is relieved and Niko also breathes a sigh of relief: "That's not a problem for us, we have our hands full here at least until May."

Gudrun adds with a smile: "We work with honey bees."

Gila is apparently not sure if she needs to understand that. Fritz and Avram Selikowitsch come into the office, where things are gradually getting cramped.

Gudrun can't resist it: "Thank you very much, Fritz, for this adventure. We don't even have to go to prison. Gila lets us get away with it once again. So please no more shelters or other military facilities today." Gila can't quite suppress her laughter: "Yes, the old guard. Always good for an adventure. Now goodbye for today. Unless you want to take a look at our apiary?" She smiles: "That's not forbidden!"

Fritz interjects: "No thanks, that's nice, I still have important appointments."

He knows when the time is right to leave. Avram Selikowitsch hugs Gudrun and Niko once again to say goodbye. They are not sure whether he has understood the situation he has put them in. They are sitting in the VW and driving up and away. Only when they have left the kibbutz grounds again does the conversation get going again. Fritz is contrite: "I had no idea that bomb shelters should not be shown to foreigners." "Yes, too bad, we were almost arrested. Thanks to Gila's flexible attitude, nothing else happened for the time being."

Niko remembers Rabinski's flexible solution when they arrived: "It was still an impressive lesson. We should have read the fine print better when entering the country. On the other hand, I can't remember a signature on a form at all. Maybe that was with the more than 20 pages of the visa application. It doesn't matter. I understand the problem here. While terror and counter-terrorism tend to be swept under the carpet in the media, this is part of daily life here. Just yesterday rocket fire? Fritz, not even you knew about it, did you? The weeping old Selikowitsch doesn't know how or when it will end!"

Fritz is silent. Of course, he doesn't know how it will end either.

They drive along the shore of the Sea of Galilee, back to Tiberias, in sunshine and in the magnificent landscape. The events in En Gev slowly fade. They are suddenly back in a seemingly ideal world full of colorful buses with Jewish and Christian tourists who want to visit the places they have heard about as children. There the dark bunker, where survival is at stake, here the exuberant hustle and bustle. The contrast could not be sharper. The Golan Heights, standing out in the haze on the other side of the lake, pose a threatening quality. Rockets that fly to En Gev also fly over the lake. Fritz is the first to regain his good mood and suggests making a short detour into the mountains on the way to Nazareth. He wants to drive via Safet, an artists' town, to a vantage point in the mountains that he desires to show them. There is indeed an artists' market in Safet. Colorful and abstract images, sculptures, stone sculptures, fashion, textiles and Schmonzes, Gudrun thinks. Many tourist buses are parked on the side of the road. Too many for a real buying mood to arise, so after a short tour they quickly drive on. The road goes steeply and high up Fritz stops at a parking lot. They get out and walk to a small wall on the slope. Their gaze falls on the Sea of Galilee, which seems to be almost within reach below them. From Capernaum to the exit of the Jordan River from the lake and the background mountains in Syria and Jordan, they can see many details thanks to the good visibility. On the other side the view is of the green tops of the pine forests to the horizon, a blue narrow band: the Mediterranean. Gudrun and Niko are deeply moved.

"One of the most beautiful places in the world, which we call Mizpeh Hajamim, which means: view of the seas."

Fritz's face becomes serious, and after a long pause he says: "If you ask me: I would swap places with the forests of the Black Forest in a heartbeat!"

Gudrun and Niko exchange glances. Was Fritz ever in the Black Forest? Obviously, the morning experiences in En Gev weigh on him more than he let on.

Thirteenth chapter

Gudrun and Niko start their experiments

The work plan for Gudrun that agreed upon with Professor Ruttner is primarily aimed at obtaining the seminal fluid in which the drones' sperm are stored in the spermatheca. Therefore, they started the production of queens without delay. In November, the weather conditions in Zrifin are not favorable. Only with the beginning of the avocado blooming in December do the bee colonies regularly produce queens. At last! The mated queens are killed and routinely prepared by Gudrun. Two small incisions with the fine insect scissors and the spermatheca is exposed with pointed tweezers. It is cleaned and pierced with an extremely thin glass capillary to absorb the spermathecal fluid. The organ is only the size of a pinhead. All procedures are carried out under the microscope and require skill and a steady hand. The filled capillary is heat sealed at the ends and frozen. Ready!

There is time between the hatching dates of the individual series of queen cells, which Gudrun uses to prepare further experimental techniques. In previous publications, the spermathecal complex has always been examined as a unit. There is a large overlying gland. Does its secretion contribute to the sperm survival? How important is the oxygenation through the dense tracheal network that surrounds the spermatheca. Gudrun makes the plan to separate the two organs from the spermatheca in the living queen and investigate whether the sperm survive afterwards. She informs Professor Ruttner of her plans.

She also discusses with Mrs. Alpern her plans. She is quickly convinced, even enthusiastic about the idea: "Can you succeed? The queens will certainly not survive." "If you don't dare, you don't win, my grandma has said!" Mrs. Alpern laughs and gathers the apparatus for instrumental insemination and accessories.

Before Ruttner's answer to Gudrun's suggestions arrives, she gets 20 "discarded" queens from Mrs. Alpern and Gudrun starts the experiment at once. She plans to investigate the significance of the spermathecal complex in separate steps. How important is the oxygenation through the dense tracheal network for the survival or fertilization ability of the sperm? Does the secretion of the large overlying spermathecal gland contribute to the sperm survival?

She plans to remove the tracheal net surrounding the spermatheca to determine what impact this will have on the spermatozoa. In a second series of experiments she plans to remove the large spermathecal gland from living queens and waits to determine if the spermatazoa survive.

In order to fix the mated queens in a horizontal position, Gudrun modified the insemination apparatus. Under CO_2 anesthesia, the abdomen is slightly elongated with tweezers. Then an opening is cut in the hard chitin part of the penultimate tergite. It has three open sides and one solid side like a flap (Fig. 12). The "window" then is held open with another hook. The spermatheca must be located directly under the opening. With light pressure and after carefully moving pushing the fecal bladder (K) to the side the spermatheca lays in the opening (Fig. 13). At this stage up to 80% of the tracheal mesh can be removed without injuring the gland. In a second series either one or both branches of the gland are removed. After lifting the hooks, the chitin flap closes the opening precisely (Fig. 14) and the queen after recovering is immediately returned to her workers.

The first operated queens lose too much body fluid during the operation and die shortly after recovering from the anesthesia. Then Gudrun finds the optimal position of the surgical window and significantly shortens the length of the procedure. The queens survive for several days and, when placed in small bee colonies, they continue to lay fertilized eggs. But two weeks after the removal of the spermatheca gland, the queens lay only unfertilized eggs. The spermatozoa can then obviously no longer fertilize eggs though it is still motile. By removing 50–70% of the tracheal net in mated queens the spermatozoa grew motionless after about three weeks.

Finally, Professor Ruttner's answer arrives. He strongly advises against the operation of queens: Queens are far too sensitive and die quickly! Gudrun laughs: "Professors are sometimes not infallible! It would be good if they recognized that too."

For Niko, it's about watching the queen lay eggs. To his disappointment, no clear differences can be found. The queen's behavior is always the same, regardless of whether the egg is placed in a worker or drone cell. First, the queen reaches into the cell with her head and front legs to inspect it, then she pulls everything out again, turns around and lays an egg. For some reason, inspection is important. What would happen if the inspection was prevented? What eggs would then be laid? Could a small transverse ridge on the queen's head prevent its insertion into the cell opening? Which material is suitable for this? After a sleepless night, enlightenment comes while brushing his teeth: The bristles of Niko's old Bockenheim toothbrush are light and stiff at the same time. Niko glues one bristle across the queen's forehead and puts her back into the colony. He is watching with excitement what happens. Gudrun is also curious. First, the queen

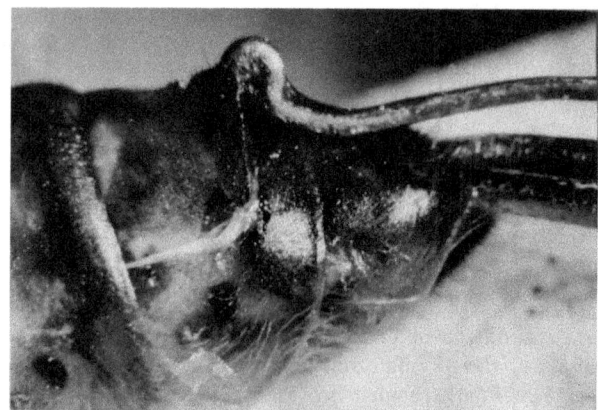

Fig. 12. The queen bee is clamped in an insemination apparatus and anesthetized. The abdomen is stretched with tweezers. In the penultimate tergite, three incisions are made with a sharp scalpel, so that this part can be opened like a window.

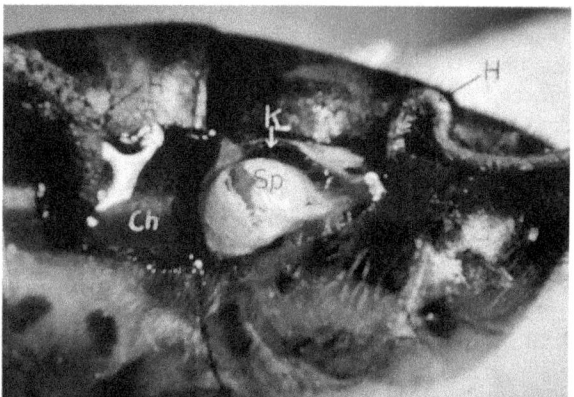

Fig. 13. The window is held open with hooks (H) and the abdomen is squeezed slightly. The fecal bladder (K) is carefully pushed to the side and the spermatheca (Sp) with its tracheal net is clearly visible. Here, the removal of the tracheal network has begun.

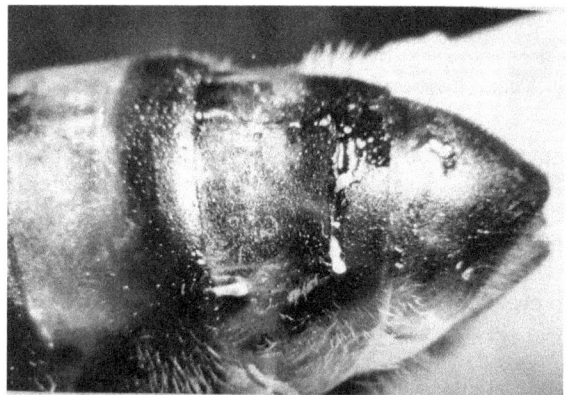

Fig. 14. After removing most of the tracheal mesh, the hooks and tweezers are removed and the chitinous flap folds back and closes the wound.

is fed and licked. Then several bees grab the bristle with their mandibles and tug at it. More and more bees participate and pull. The bristle with the glued queen is moved towards the hive entrance. The workers become aggressive and even try to sting the queen. Fifteen long minutes afterwards—to the relief of the two observers—the glue comes off the queen's head and the bristle is carried through the entrance. The behavior of the workers quickly normalizes, and just ten minutes later the now unhindered queen lays her first egg.

His first results: Workers don't like queens with toothbrush bristles on their heads. Niko has to buy a new toothbrush, and a new plan has to be found: He now glues small flags made of plastic film on the queen's front legs so that she can no longer insert them into the cell.

After feeding and licking the queen, individual bees try to grab the flags with the mandibles and pull at them. Another failure? But no! This time, the queen defends herself with her middle legs. Egg laying occurs without the queen being able to inspect the cell. Niko is eagerly awaiting further developments. To his great disappointment, these eggs result in normal worker bees. The prevention of inspection by flags has not changed anything. A big disappointment! Fertilized eggs have been laid in the worker cells, from which worker bees have developed.

What happens when eggs are laid in drone cells? Niko locks the queen with attached flags on combs with drone cells. She also lays eggs in the large drone cells. Now it's time to wait long days again. Then the big surprise: Only worker bees emerge, not drones. A great result that allows only one conclusion: the queen measures the size of the cell with her front legs and then lays either a fertilized or an unfertilized egg. The mechanism seems to have been found.

A few days earlier, an overdue aerogram from Professor Ruttner had arrived, in which he strongly advised against experiments on drone combs as long as the techniques to obstruct the inspection were not better developed: "This is a premature experimental dead end". They send the report on their results to Oberursel full of enthusiasm, but not without concerns. How will the Professor react to their results, which run counter to his advice? Professor Ruttner's care "suffers" or, as Gudrun mockingly notes, "wins" due to the long delays caused by airmail. The airmail letters with their questions from Israel take ten days to reach Oberursel. Even if Professor Ruttner answers immediately, they will not receive his advice for 20 days at the earliest. Therefore, they cannot wait for a reply from Oberursel. Gudrun and Niko have to take over the further development of the experimental program independently. First of all, there may be accusations because of the unauthorized action. To their delight Professor Ruttner is obviously pleased with the results and progress. He doesn't skimp on praise and recognition.

Fourteenth chapter

Niko gets an operation in the military hospital

Winter is also coming to Israel. In January and February, the temperature drops to a few degrees above zero at night, and only 10 -15°C during the day. There is no heat in the house. Gudrun and Niko put on all their clothes and still freeze. Eventually, Gudrun starts coughing and develops bronchitis. Niko drives her to Bela's pharmacy in Ramla, who listens to Gudrun's cough: "That doesn't sound good." He looks worried: "I've got just the right thing" and pushes a box of tablets and a bottle of cough syrup over the counter. "What do we owe you?" Gudrun croaks.

Fig.15. The queen with small flags made of plastic film glued on the queen's front legs so that she can no longer insert them into the cell.

"This is my gift for you, Gudrun, get well quickly!" answers Bela.

"We can't accept that!"

"Yes, that's okay. We humans have to help other humans, no one else helps us."

The two are surprised again. This would certainly not have happened to them in the pharmacy at Bockenheimer Warte in Frankfurt. The cough subsides and after three days Gudrun is healthy again!

Niko is the next to fall ill. For several days, he cannot keep any food down. He quickly weakens, which he denies. Gudrun observes that he has difficulty bending down and tying his shoelaces. Stomach ache? Soon Gudrun can no longer move in bed without Niko moaning in pain. She suspects appendicitis, which she had as a child years ago. Again, on to Bela in Ramla. He certainly knows what to do. Niko manages to start the Horex as usual at the first kickstart, but swinging his leg over the bench is difficult. The pharmacist looks at Niko seriously and talks to a doctor on the phone. Only a quarter of an hour later, Niko is examined. Gudrun's diagnosis is confirmed by the doctor: acute appendicitis! A telephone conversation from the doctor leads to admission to the military hospital neighboring the honey bee station. They have to go there themselves. Gudrun's attempt to start the Horex fails. She cannot press down hard enough to push the kick starter of the large machine. Cold sweat beads from Niko's forehead. He starts the Horex with his face contorted in pain. They drive home, pack Niko's things and, with Gudrun's help, creep through a hole in the fence of the plantation to the military hospital. Everything is prepared there, the doctor had been called. Niko is laid on a mobile couch and is not allowed to take another step. A young blonde ward doctor greets them in German. That is reassuring. Later they learn that the doctor comes from Berlin. She examines Niko and orders immediate surgery. Without delay, he is driven to the operating room for anesthesia. Gudrun has to wait in the corridor outside the door. The operation does not take long. The doctor comes to the waiting Gudrun, beams and points with both hands: "Such a thick appendix! He was on the verge of it bursting!"

Niko is still dazed from the anesthesia. Nevertheless, he recognizes Gudrun, wants to stand up and hug her. Gudrun quickly hugs him and pushes him back onto the bed. He is taken to a large hospital room with 20 beds. There are 17 patients. All young men in green pajamas. Several family members sit at each bed. A lively atmosphere—very different from that in German hospitals. Niko falls asleep again. He will put on the green pajamas later. Gudrun has to go back to the station and take care of her queens. Everyone there wants to hear how Niko is doing.

The next morning, Gudrun asks Mrs. Alpern to call the hospital and ask how Niko is doing after the first night. "He's fine, but everyone there asks why you're not with him yet!" "I inquired about the visiting hours. It said: "From 4 to 7 p.m." Mrs. Alpern laughs: "Nobody adheres to such regulations here." "Not even in a military hospital?" asks Gudrun. She shakes her head in amazement. She hurries back, quickly takes care of her queens and checks whether Niko's bees are okay. Just an hour later, Gudrun is standing in front of Niko's bed. He is lively and happy to see her. Some other people seem to be happy as well. That a patient has undergone surgery without relatives has caused great concern. A nurse asks why she is only coming now. Visiting hours? The nurse doesn't understand that. She brings Gudrun a hot cocoa so that she stays there! For the other patients, it's the other way around, there are food basket everywhere, brought by the visitors. In the afternoon, Gudrun has to go back to take care of her and Niko's bees. At the station, Meir waits for news about Niko. One of his sons works as a nurse in the hospital and has told him a lot. The operation was not easy because the appendix was severely swollen: "Oiwawoi, poor Niko! He must be suffering so much, Oiwawoi!"

Meir throws his hands up in disbelief. You can see how he suffers. Later, he comes to the house for dinner, not without his wife's deliciously prepared food.

"Today I can't eat anything," he complains. "Poor Niko is not allowed to eat anything either. He's certainly in a lot of pain." Meir has to pause moaning in order to shove a few heaped spoons into his mouth. "I don't like the food today, poor Niko."

Another break. He notices that Gudrun is eating quietly: "I thought you loved Niko! How can you just eat there! Poor Niko must be very hungry!" Gudrun has now developed an understanding of the oriental way of compassion, despite everything she does not succeed in expressing her feelings so dramatically. Meir's request to visit Niko is rejected by her: "It's much too late, they won't let us into the hospital."

Meir does not give up. His son is currently on a late shift and his father is also known in the hospital. Meir takes his old flashlight and slips through the fence to the hospital with Gudrun. Gudrun is surprised: the front door is open. She finds the way to the hospital room easily. Here, too, the door is wide open and, as always, there is a lively hustle and bustle. No sign of a night's peace. That's when Meir discovered Niko. With loud lamentations of "Oiwawoi" he throws himself on Niko's bed. Tears run down his face: "I can't sleep or eat because I've been so worried about you. Thank God you didn't die after all!"

Niko's face contorted in pain indicating that Meir must have hit on the fresh surgical wound during his hug. Then Niko turns to the side and assures him that he has been waiting for Meir's visit for a long time.

"That's great, Meir! Despite your many duties and work, you have found time to visit me. This is an immense joy for me, a very important visit, and a true honor."

Meir and Niko exchange the nicest courtesies and assure each other of mutual appreciation and friendship. Gudrun, on the other hand, only asks succinctly: "Niko, how are you?"

After half an hour, Meir says goodbye, complaining loudly and among numerous other "Oiwawois", because his duty is calling. To his regret, he could not spend the whole night at Niko's bedside, which of course he would prefer to do. He has to water the plantation. Gudrun gives Niko a quick goodnight kiss, mumbles something about getting well soon and runs after Meir.

The next day, Gudrun gets up early. She takes care of the queens and Niko's bee colonies and is in front of Niko's bed before ten o'clock. Niko is doing much better. His face is no longer pale, it shows the usual suntanned color again.

"Good Meir! That was an experience last night that I will not forget! Many people in our room asked me in amazement if Meir was my father. I initially confirmed this. Then I told the truth! Everyone then wanted to know why we are "so closely" friends with Meir." He laughs.

"Resting and sleeping is not the order of the day here. I was still very tired when a physiotherapist woke me up this morning. She tormented me quite a bit with "One, two, three". Getting up and "keep moving" are the top priorities here."

Then Niko looks embarrassed: "Do you know what happened to me this morning?"

"How am I supposed to know?" answers Gudrun, slightly irritated.

"On the way back from the toilet, I fainted. Anyway, I was on the floor. When I came to consciousness again, there were many curious faces above me. It took me a short while to understand the situation. My pajama pants were way too wide and had slipped down to below my knees. Many people have probably not seen a man who has not been circumcised for the first time." For Gudrun, this sounded pretty embarrassing. Niko is not squeamish. He feels quite comfortable in the hustle and bustle of patients and visitors and has made friends with some women soldiers from the next room, in blue pajamas, of course.

When Gudrun returned back to the bee experiments, the soldiers begin to question Niko about his military service as a pioneer in the Bundeswehr. He answered the many questions as best he can. In return, Niko asked about the long three-year conscription, but received only tight-lipped answers. Obviously, all soldiers are effectively "trained" to say nothing about the Israeli army. Niko asked and it was confirmed that military matters should never be discussed, not even in a parents' home.

Niko was very surprised by the great interest of some Israeli soldiers about the German "Blitzkrieg": "How did the German tanks overcome the Maginot Line? What was Guderian's tactics?"

Niko was also asked about the successes of General Rommel's tank army in North Africa. "I've heard the names of General Erwin Rommel and Heinz Guderian before, but I really don't know how and where their tanks advanced. Actually, I'm not interested in that either! Only the incorrigible militarists are interested in that at home!"

Niko is angry. He does not understand why the young generation of Israel, whose parents suffered immeasurably under German militarism and National Socialism, is so blithely enthusiastic about the initial German war successes. Only later does he realize that a drastic change of role has taken place with this new generation. How it is important to defend themselves and always be the stronger. Under no circumstances should one become defenseless victims again!

While Niko is in the hospital, many people feel sorry for Gudrun, who now lives alone in the house with the bee colonies on the large plantation. The concern for Niko is also clearly noticeable. The bee people come by more often and offer their help. Everyone is surprised when Niko reappears just three days after the operation. Gudrun, on the other hand, is not surprised. Typical Niko, she thinks. However, she is amazed that the doctors agreed to such a quick discharge. Twelve years ago, she had to stay in hospital for ten days after her appendectomy. A few days later, Niko goes back to the military hospital to remove stitches, but only for a few hours.

Fifteenth chapter

Gudrun and Niko harvest honey, say goodbye to Horex and Israel

The orange blossom season has begun. The land is enveloped in an intense, sweet fragrance. On their weekend trip to Dwora and Fritz Kuhn's home, they virtually bathe in this sweet floral scent. The bees are also delighted. They fly from the first dawn until late in the evening to the blossoming orange trees. However, the nectar that is collected interferes with the production of queens. In addition, the fresh, not yet completely healed wound from Niko's appendectomy prohibits lifting the heavy supers of the colonies, which are increasingly being filled with honey. Mrs. Alpern sends Mr. Sushinski. Isn't he too old for that? When they cautiously asked whether he can take over the job, he reacts extremely indignantly: "What's the point! I'm strong and I've proven that many times. Recently, the ministry sent an official. They wanted to dismiss me because of my age. I need work, otherwise I'll die. I asked the young man from the ministry to wrestle with me. If he wins, I will accept my dismissal. If I win, I can continue to work. Of course, he was afraid. A coward. He didn't fight and I can continue to work. Now you doubt that I can lift these few honey supers?" Niko also doesn't want to fight with Mr. Sushinski, which Gudrun can understand. They are surprised at the ease with which old Sushinski takes down and puts on the heavy and full honey supers. He now helps them regularly and talks about his experiences as a partisan in the fight against the Wehrmacht in the forests of Eastern Europe. What a will to survive, even optimism, can be heard in these reports! How and when did he come to Israel? With Moledet as DP? He pauses, his face hardens to stone. Grey Ashen! He speaks facing the heavy super, which he is lifting without looking at it. "1945, to Treblinka." Then he continues to work in silence. Only slowly does life return to Sushinski's face. Mr. Sushinski does not want to and cannot say "anything" more. Dismayed, they remain silent. There are stories that cannot be told. Especially not to Germans.

The large amounts of honey that their bees have collected fill many cells in the brood nest of the colonies and hinder brood rearing. For Niko's experiments, the queens need space and empty cells for egg laying. They have to get the honey out. Mrs. Alpern offers to lend them a honey extractor on Shabbat.

Now it's time to get down to business! Even before the Shabbat rest begins on Friday, they quickly buy seven large buckets and some large screw-top jars in Ramla. After that, they work until late at night. They have to return the extractor on Sunday. They harvest more than 200 kilograms of orange honey from their four colonies. The next morning, the floor and the large table are nicely patterned with small and larger black rings. Only a closer inspection can explain that miracle: Each drop of honey is surrounded by a dense row of black little ants! They heat water, then wipe thoroughly. Ants are everywhere, albeit less. The procedure has to be repeated again until the last honey sources for the ants have dried up. Fortunately, the floor is made of stone slabs and not of wooden planks as in the Oberursel honey house.

On Sunday, Gudrun and Niko visit Mr. Hiller with a jar of honey. They ask him to tell the workers to bring jars which they want to fill with honey. A small "thank you" for the help at their arrival. Of course, the other friends, Meir, Bela in Ramla, the Kuhns in Pardess Hanna and many more are also given honey. Even after that, there is enough for her own breakfast. Freshly squeezed fruit juice, orange honey on flatbread and a hot cup of Nescafé: bee research in Israel can be so beautiful!

Their plans then undergo an unexpected change. Professor Lindauer offers Niko a chance to become the replacement for an assistant position. However, Niko had to sign his employment contract before May 15, 1967. What should they do? They have no choice. Professor Ruttner sends them flight tickets for May 10. They can't refuse. They find it important to bring the ongoing experiments to a quick end. The main thing now is to carry out as many repetitions of the experiments as possible. The experiments are now running like clockwork. The results of Niko and also Gudrun's surgical experiments agree very well with previous results: If queens are hindered to inspect the brood cells before egg laying, they lay worker eggs into drone cells. Gudrun can show, that two weeks after removal of the spermathecal gland the queens produce only drone brood. When checking the spermatoza in the spermatheca it is still motile. That proved to be different after removing the tracheal net but leaving the gland. Then the spermatoza lost its mobility. The test repetitions can be statistically confirmed. At last, a solid foundation for their dissertations. Also, she has many glass tubes with spermathecal fluid from more than 100 queens.

They had arrived in Israel with only a backpack and two side bags. Now there is much more to transport. Just the test protocols and the samples. In addition, there are the utensils they have purchased. Again, Dwora and Fritz Kuhn help. They give them two large suitcases, which they fill to the brim with their things. How many things did they get and buy during their stay in Zrifin!

Another problem is the motorcycle. At first, they thought they could simply sell the Horex. The mechanic of the motorcycle workshop in Ramla offered a quite acceptable price. Mr. Hiller warns that the machine was not cleared through customs when it was imported. So, if it is now introduced, customs duties will have to be paid. The information from the customs office is sad: the import duty far exceeds the value of the motorcycle. A return transport would also cost them dearly. There is no alternative. The Horex must be handed in at the customs office. This is the only way they can leave the country legally.

On May 10, Fritz Kuhn picks them up early in the morning by car and stores the two large suitcases in the back. While Gudrun can go with Fritz by car, Niko follows on the Horex. They drive to Meir in Ramla, who wants to accompany them to the airport. At the main customs office at the airport, they have to leave the Horex with a heavy heart. The motorcycle gets a seal from customs and the papers have to be handed in. Niko takes a deep breath as the customs officials hand over the receipt of the machine. So many memories of the countless tours all over Europe to Israel. The Horex has been an important sympathetic figure for them. Over and gone! A last clapper on the driver's saddle and in Fritz Kuhn's Beetle to the terminal.

The airport of Lod is small and cramped. Because Lufthansa has no landing rights in Israel, KLM has to take a detour via Amsterdam. At the check-in counter, the stewardess points out that her suitcases weigh more than 55 kilograms together. Only 40 kilograms of free baggage allowance is permitted. The airline is happy to transport extra luggage, at a cost of more than 100 dollars. They don't have that much money. No other alternative: open your suitcase and get out the unnecessary. They block the counter. Dirty laundry, the bee overalls, rubber boots and the rest of their equipment do not meet the high standard of distinguished air travelers.

Fritz, who attaches great importance to appearance and level, stays discreetly in the background and acts as if he has nothing to do with them! Meir helps them unpack and finds the situation in no way embarrassing. He is happy about the rubber boots, jackets and other old things. They pack up the most important things again. Still six kilograms too much! The stewardesses are getting nervous. Some passengers behind them complain. After they reduce the excess baggage fees to twelve marks, about three dollars, they refrain from unpacking again and check in their suitcases. Meir gets himself strings and packs the "Alte Sachen" into two bundles, which Fritz stows in his car with some reluctance. Nevertheless, his farewell is very warm. His eyes are red. Meir cries big tears and complains loudly that their departure plunges him into deep grief. He will miss them forever. Gudrun and Niko are also sad, and they will never forget him.

When they leave in tears, they have no idea that the Six-Day War will break out just four weeks later. The remilitarization of the Sinai by Egyptian troops is in full swing as they walk up the gangway to the plane. On May 15, 1967, Israel's Independence Day, which is celebrated with a military parade, the Egyptian armed forces take up their positions. The UN troops controlling the situation on the Egyptian border are completely surprised. Jordan and Syria are joining the attack plan. The war begins on June 5, when Niko supervises his first course in Frankfurt. A surprise attack by the Israeli Air Force on all Egyptian airfields ensures Israel's success. The ground troops ensure a rapid military victory. Niko has difficulties to finish the internship objectively. Niko remembers the questions of the soldiers in the hospital. Apparently, they were prepared for this war in the most precise way. What would have happened if Gudrun and Niko had returned by the Moledet as planned? What would have become of Israel and them if the Israelis had not been so well prepared for this armed conflict?

Part 2

Final sprint to the doctorate
Mother's happiness and father's career

Sixteenth Chapter

Gudrun and Niko settle back in Frankfurt and greet an aunt from America

After their return Gudrun moves in to Niko's room in Seestraße. The landlords warned them:

"A young couple like you in such a small space? That won't go well? That's not even 12 square meters."

Mr. Werner has been married for 40 years and knows what he is talking about. Niko knows from Friedrich Schiller, whom he otherwise appreciates rather little, that "there is room in the smallest hut, for a happily loving couple."

And now it's time for the researchers to complete their dissertations. As soon as there is more time, they can take care of their "housing problem". Despite her wedding costume, Gudrun does not have a particularly lavish wardrobe, and so Niko's wardrobe is divided fairly. Fair, not equal: Gudrun two thirds, Niko not quite a third. The shelf made of bricks and boards is expanded a little, and Gudrun's books fit in.

For his employment as a representative of an assistant position, Niko quickly learns that he is back in the land of orderly bureaucracy. It is not enough to sign a contract. Several appointments with the university administration are necessary. An income tax card must be obtained, registration for unemployment insurance, pension insurance, health insurance and a police clearance certificate. Niko hates bureaucrats. He grudgingly surrenders to the bureaucratic paper war. At the end, his four hours of teaching per week are actually twelve, and they have to be thoroughly prepared. Niko does not want to embarrass himself with the students. He also stands in for Professor Ruttner when he is unable to attend, and he is absent more often than Niko would like. Gudrun helps with the organization and supervision of Niko's courses. In any case, the teaching tasks take much more time than he had thought.

For the daily trips to the Bee Institute in Oberursel, Gudrun and Niko, without the Horex, are dependent on public transport. They quickly realize that two monthly tickets are far too expensive. In addition, the tram is infinitely slow and they often have to commute between the university and the Bee Institute several times in one day. A new vehicle is needed! The financial leeway has increased. Gudrun is pleased to note that the Zoological Institute continued to pay her small salary as a student assistant during her time in Israel. She had completely forgotten about this—and is now unexpectedly noticeable in the account in the form of 2400 marks. Together with Niko's salary, they could now actually afford a car.

The list price for a new VW Beetle is DM 4,485.

"That's more than I thought," Gudrun points out. "Or do you prefer a used one?" "No, absolutely not. With the Horex, the motor was clear and so simple. You could see everything. In the Beetle, the engine is hidden in the back and is difficult to access! We need a new car."

A branch of VW Glöckler is only a ten-minute walk from home. In the sales room they meet Mrs. Riese.

"Recently, our bike broke down. A repair is not worthwhile. We want to buy a new VW Beetle and have 3000 German marks!"

"I have a suggestion especially for you." Ms. Riese smiles and takes off her golden glasses. "We still have a few new cars from the 1966 model. They are available immediately, no delivery delays as with the current model 1967. Three thousand marks is too little even for the previous model. Take a loan of 1000 marks, and the matter will be done today." Mrs. Riese nods at them encouragingly.

"A loan as a student without a regular income? How is that supposed to work? No, even 4000 marks for a pre-model is not possible for us! Please come a little closer to us," explains Gudrun in a very friendly tone.

"Young woman, you impress me. You don't give up easily! Maybe I can reduce the price to DM 200. That would bring us to 3,800 cash. No further discount."

Gudrun nods and asks: "That sounds better. Niko, what do you say if I increase our offer of 3000 by 200 as well?"

Mrs. Riese seems annoyed: "We should come to the end. I have an appointment in 30 minutes The difference between 3,800 and 3,200 is exactly 600. If we divide that, the car costs 3,500. This is my very last price! Hop or Flop! I have been negotiating with you for longer than the matter is worth to me!"

"Well, and thank you very much, Mrs. Riese! When will the car be registered?"

Three days later, the new lotus-white Beetle Model 1966 is parked directly under her window! Niko and Gudrun are relieved. They saved more than an hour per day on the journeys from the university to the Bee Institute and back. Time that they can use well!

Now that they are proud car owners, social obligations arise. Niko receives a letter from his parents in which they tell him that Aunt Louise from Philadelphia is planning a visit to Germany. She will land at the airport in Frankfurt. Please pick her up with the Beetle!

"Who is Aunt Louise?", Gudrun wants to know.

"One of my grandfather's brothers emigrated to the United States before World War II. His daughter Louise, who was called Aunt "Uliese" by we children, regularly sent us care packages in the post-war years. It was always a big party. From peanut butter, Cadbury chocolate to chewing gum, everything was there!"

"We didn't have a real aunt, but we still had an American family who regularly sent us care packages. It was like Christmas for us every time. Unfortunately, we never got to know this family."

Niko and Gudrun are curious about Aunt Louise. They know America mainly from Hollywood movies. An unknown continent with unlimited possibilities? What will Aunt Louise be like? Like Audrey Hepburn? Or more like Doris Day? Or even glamorous like Marilyn Monroe? They quickly realize for themselves how unrealistic their image of

America must be.

One hour before arrival, they make their way to Rhine-Main Airport. Gudrun is waiting in the short-term parking zone. Niko has armed himself with a large sign "Aunt Louise" and takes up a position at the arrival gate. The display board announces that Pan American Airlines Flight No. PAA 72 from New York will land punctually at 8:15 a.m. When travelers come through the exit door, he holds his sign particularly high up. Finally, an elegantly dressed older lady, clearly of the Doris Day type, comes confidently through the door. Niko stretches. She looks around searchingly, her gaze lingers on Niko's shield —and her white teeth contrast strikingly against the dark red lipstick. Beaming, she marches towards Niko and is not noticeably slowed down by a huge blue suitcase on wheels, the color of which matches her light coat. Chic aunt – Niko is impressed. He takes down his sign when he is sure that he has been recognized and runs towards her. Even before he says: "I'm Niko. Welcome to Frankfurt", Aunt Louise drops her suitcase and hugs him.

"Your parents wrote so much about you and your beautiful young wife. Can I meet Gudrun?" "Of course, she waits in the car right at the exit from the terminal. Come on, I'll take your luggage."

Niko takes the big suitcase from her—which is even heavier than it looks—and they move through the terminal together. On the way, Aunt Louise talks non-stop, her German toned and heavily mixed with American vocabulary. Because of the fast flow of speech, Niko can at best answer every second or third question. That doesn't bother her.

Gudrun's greeting is just as warm, almost more effusive than Niko's and is accompanied by many compliments. Then a problem: the big blue suitcase doesn't fit in the trunk. Aunt Louise is irritated that they want to put it in the engine compartment and points to the rear: "Won't it fit there?"

"No, there's the motor," explains Gudrun. In any case, the dimensions of the luggage are not adapted to a German standard VW beetle. Niko suggests an alternative: "Maybe the suitcase fits upright on the passenger seat."

In the end, after a few tests, the occupancy plan is perfect: Aunt Louise – somewhat hindered by her tight skirt – skillfully juggles her way into the seat next to Niko in the back seat. Gudrun drives, a little trapped by the suitcase. Aunt Louise talks, and talks, and talks. Niko nods, and nods, and nods. Gudrun drives, and drives, and drives the VW Beetle. Only the suitcase is out of the ordinary.

In Bockenheim, they eat with Aunt Louise at Salvatore's, the Italian restaurant on the corner opposite to their room. They don't want to shock Aunt Louise even more with Hessian "Handkäs with music" and sour cider. Pizza is – as far as they know – a staple food in the USA. The calculation works out and Aunt Louise is probably very hungry and enjoys her pizza Hawaii. Only now and then a short "lovely" between two bites. The broken flow of speech allows Niko to explain the further planning of their trip to Bad Salzuflen. Apparently, his parents have not yet informed Aunt Louise in detail: "In the afternoon we will drive you to the main railway station, there you can take the train to Bielefeld." Aunt Louise seems happily surprised: "Oh, by train! Not in the Volkswagen? Okay, that's fine with me. Will it take a long time?"

"Your train leaves at half past four from Frankfurt to Bielefeld without changing trains. My parents will pick you up at the train station there."

"That sounds perfect. There's still time until then." She looks at her watch. "You have to show me your apartment first."

Gudrun and Niko look at each other in surprise, No problem: the room is reasonably tidy and the sofa bed is folded in. Aunt Louise takes over the bill, and neither of them has a problem with that either: "Thank you very much!" They are sitting in the same formation in the Beetle and driving around the corner to their room in the "Kleine Seestraße". The landlady, Mrs. Werner, waves friendly and is surprised by the transatlantic whirlwind that Aunt Louise creates. Doris Day in the Kleine Seestraße, wow! The poodle has hidden somewhere to be on the safe side, and is barking loudly in the background. No trace of Molles, the cat. With loud shouts and many oohs and ahs, they land in front of the ladder that leads up to the room.

"Up there?"

Without hesitation, Aunt Louise takes off her high heels. This is where her meteoric career with the "Girl Scouts" in Philadelphia comes into play. Elegant and nimble, she climbs up. Once at the top, she visits the facility and—yes, that is also possible!—is speechless for a moment.

Then she laughs loudly and as bright as a bell:
"You live here? And you sleep on this narrow couch every night?"

"No, no. It can also be wide."

Gudrun demonstrates the pull-out mechanism, which expands the lying surface to a width of a whopping 100 centimeters. Even that causes Aunt Louise to be even more astonished. Her bed, in which she sleeps alone, is queen size—much bigger! "Do you know the TV series "Love on the roof top" with Judy Carne and Peter Deuel? I watch them every week. Hilarious! I absolutely have to suggest to the film team to come by here. Then they would see what living together in a confined space really means. They don't even know what's really funny."

Gudrun doesn't know the soap series, finds it only funny to a limited extent and points to the advanced time. Aunt Louise would like to stay longer. The train doesn't wait.

A few weeks after Aunt Louise's return to the USA, an airmail letter arrives with her heartfelt thanks for the beautiful hours in Frankfurt. She still has a question: "Where is your bathroom?"

Well, they could answer that they don't have a bathroom and therefore visit Aunt Anna in Frankfurt Sachsenhausen once a week, every Tuesday evening. There they enjoy a large bathroom, the washing machine and a sumptuous dinner on top of that. They think that would be too complicated—so Aunt Louise waits in vain for their reply. After all, the questions seemed more important to her than the answers anyway.

Seventeenth chapter

Gudrun and Niko fight for bees for their experiments

The Institute is lively. Everyone tries to implement as much of their program as possible. In addition to their own work, Gudrun and Niko also take part in the projects of the other candidates. Mr. Berrmann has drone brood combs in the incubator and collects the drones every day. They are marked with different colors and introduced in colonies. From there, they are removed after different times and preserved for histological examinations. He wants to examine the development of a mandibular gland of the drones.

Gudrun and Niko think this is a boring and simple work program. They are not surprised that Berrmann is the only doctoral student at the Institute to keep regular working hours. Conversations with him also shows that he does not stand behind his project. He preserves the heads of his dated drones, but he has no answer to the question of what comes out of it: "Now I'm going to collect and preserve the drones. After the bee season, in autumn, I will do the histology and then I can tell you how the drones' mandibular gland develop."

"Is this person not curious at all?" Gudrun thinks. At the beginning of her own work, she too had only collected without checking whether the method worked. Then, when thawing at the end of the bee season, all spermathecae had burst! Since then, they have been evaluating their own experiments as quickly as possible. They want the results as soon as possible. This is the only way they can improve their technique and methods! In the fall, it becomes clear that Berrmann's project has failed. Despite months of histological examinations, and stains, he cannot find any mandibular gland. He has worked in vain for a whole bee season, Professor Ruttner is angry. Nevertheless, Berrmann is getting a new topic for his doctoral thesis. This time it will be about the queen's neurosecretion. In the future, he will immediately start the histological evaluation of his specimens.

Gudrun and Niko urgently ask Professor Ruttner for an appointment for a detailed discussion. The main focus will be on how many bee colonies will be available for their experiments. Professor Ruttner has little time and they are put off until the coming week. A delay that further shortens the advanced bee season.

"Especially for the operation, I need many queens in small nucs again," Gudrun demands. Professor Ruttner agrees and the work meeting is constructive. Gudrun has a more than full work schedule: the continuation of the surgical removal of trachea and glands of the spermatheca is on the agenda, in addition to the original topic, the chemical analysis of her liquid contents, which she collected in Israel.

After Gudrun's meeting, it's Niko's turn. He meets a tired and unfocused supervisor and reports to Gudrun:

"I gave him my experimental plan a few days ago. Obviously, he has not read it. Professor Ruttner also did not answer the question of whether I can get the 24 bee colonies I need for my experiments. Only a completely non-binding promise that he wants to talk about it with master beekeeper Blotz. And then his statement: If necessary, we just have to buy bees!"

Niko is indignant and bitterly disappointed:

"Who will sell good bee colonies before the honey harvest?"

"Calm down!", Gudrun soothes.

"You know about Professor Ruttner's many different commitments and appointments. Our work is only a minor matter for him. We have to and want to carry out our projects independently." She adds: "There will be no lack of the necessary freedom and support. Think of Israel."

Gudrun gets a good microscope and a modern insemination apparatus for queen bees. During its modification of the insemination device, the hooks for opening the sting chamber have to be replaced by fine tweezers that she got from the precision mechanic in the workshop of the Zoological Institute. This way the surgical technique is optimized. Most queens survive for many weeks in small colonies (mating boxes). As the number of surviving queens increases, so does the time required for care and monitoring. Gudrun reaches the limits of her performance as the season progresses. Of course, Niko helps. The last evening activity in the Institute for Gudrun and Niko is a tour of the garden and the joint control and care of Gudrun's experimental colonies.

Niko's experiments are good at first. After the inspection of brood cells is blocked by flags, the queens usually lay about 80 percent fertilized worker eggs in drone cells. After removing the flags, the queens no longer make mistakes. Only unfertilized drone eggs are placed in the drone cells. For these experiments, Niko used all twelve bee colonies available to him at the Institute.

"I have to move forward and finally tackle the question of whether the queen's front legs are involved in recognizing the drone cell. Nothing works without more test colonies!" he complains.

Professor Ruttner offers him a cup of coffee. He is friendly: "Yes, Niko, you still have to learn that, after this bee season comes the next bee season and then the next one! Patience, perseverance and, above all, humility are necessary if you want to study bees. Bees are small animals, and when money is to be distributed, the little ones usually get little! Many politicians and decision-makers don't know anything about bees and so we usually get our turn last. Perhaps not least when I think of our new building and honey production. And Niko, here at the Institute, everyone has to adapt his program to the limited resources we have. This applies to me as well as to you and to our beekeepers who need a new, expensive wood machine. There is also good news for you. Yesterday a Mrs. Ströbel from Darmstadt called me. Her husband, a beekeeper, is in hospital after a serious car accident and can no longer look after his twelve bee colonies. He wants to hand them over as quickly as possible. As Mrs. Ströbel says, she prefers honey to money as payment. I will be able to settle that to the satisfaction of both sides. Write down this phone number and call Mrs. Ströbel. Please talk to master beekeeper Blotz when the Institute bus and trailer are free."

Niko is happy, finally more bee colonies! Two days later, Gudrun and Niko drive towards Darmstadt in the evening. As reinforcements, they have taken Martin Dautz, a new doctoral student, with them. He wants to help loading the hopefully heavy beehives. It is not yet completely dark when Niko closes the first colony entrance with a plug. In the process, he probably squeezes individual bees. Many bees from the colony smell the fruity scent of the alarm pheromone released by the squeezed sisters—and bees of the other colonies perceive the alarm smell. They react with defense! The flashlights are the first target of the stinging flights. They quickly turn off the last lamp and climb into the bus, cursing and whining. They close windows and doors and, in the light of the vehicle's lamps, remove the bee stings from their faces, hands and other parts of the body. A painful lesson. They wait for complete darkness. Then they put on their protective suits along with face veils, rubber boots and gloves and carefully check each other for gaps in the protective clothing. Now they can safely get out and start the smoker. Using sufficient smoke, they carefully close the entrances of the other beehives without squeezing a single bee and without being stung.

The beehives are heavy. Obviously full of honey! Martin and Niko sweat in the warm protective clothing. Sweat is better than more bee stings! Gudrun leads the way, the men are slowly following behind. When six colonies are in the car, they have to take a break. Finally all 12 colonies are in the car and the trailer. They drive back satisfied. When the last bee colonies are unloaded and set up in the Institute's garden, it is shortly before midnight.

The next morning, the beekeepers and colleagues laugh at them. Professor Ruttner also smiles mockingly when he sees Gudrun and Niko and asks about yesterday's adventures. Their faces are swollen from the many bee stings, Gudrun's eyes are closed except for narrow slits. Only with large sunglasses can she avoid pitying looks from passers-by. Niko's upper lip is swollen. He cannot close his mouth, which leads to constant salivation and contributes significantly to his disfigurement. Martin Dautz's wife calls and excuses her husband for being ill. Fortunately, these problems are over the next day.

The bee colonies brought from Darmstadt, known as "Ströbel colonies" for short, pose a real challenge. The intensive defensive readiness of these bees requires complete protective equipment, bee veil, combination suit and boots, only gloves must be dispensed so that Niko can catch the queens. A lot of smoke is necessary to tolerate bee stings in the bare hands within such limits that Niko can withstand. The clouds of smoke and protective clothing bring Niko suggestions about his suitability as an astronaut on Mars, as well as hilarity and laughter in the Institute.

Finally, Niko finds a trick: he places the colony he wants to work on about ten meters away and waits until all the foragers look for their colony at the old hive location, disoriented. Without these stinging bees, the colonies are easier to manage! The Ströbel colonies enable him to clarify the function of the queen's front legs in the perception of the drone cell for the first time experimentally. After unilateral amputation of one of the queen's front legs, there is no impairment of cell recognition. Thus, even with only one front leg intact, the queen lays unfertilized eggs in drone cells. The experiments with complete amputation of the front legs on both sides, on the other hand, seem to largely prevent the perception of the drone cell. Queens without front legs prefer to lay fertilized eggs in drone cells. Although some repetitions of experiments are still necessary, the importance of the front legs is clearly apparent.

As discussed with Professor Ruttner, after Niko's experiments the Ströbel colonies are handed over to the beekeepers of the Institute in order to fill mating boxes and make artificial swarms. The Professor wants to make sure that no Ströbel queen or her daughters end up in the Institute's colonies. A few days later, when Niko and Gudrun approach the Institute, they see a large cloud of smoke over the Institute. Gudrun thinks that the Institute is on fire. Has the fire brigade arrived? As they get closer, they see the smoke coming from three bee smokers. Two beekeepers, protected only by bee hats, jump around the Ströbel colonies, cursing. Gudrun and Niko laugh.

"Come on, we'll help them," Gudrun suggests.

"We'll leave it at that. The one, who laughs last, laughs best!" says Niko with a grin.

Eighteenth chapter

Niko is visited by the MAD

Martin Dautz was given a classic topic of bee biology for his doctoral thesis—the dance language of honey bees. Professor Ruttner wants to position his group thematically broad. Although all of them are to work on honey bees, the entire breadth of biology is to be investigated. Martin is supposed to find out how well foragers can recruit other bees in the hive to a food source. The basic mechanism has been known for a long time. The successful forager communicates the location and quality of a food source to forager bees on the comb via dance language. How many dance circles must be followed before a follower bee successfully finds the indicated food source? Is one dance enough or does the follower have to follow several dances before she arrives at the right place? Martin has developed a good plan and explained it in detail to Professor Ruttner.

Now his experiment starts and in a short time more than 1000 newly hatched workers have to be marked! For this purpose, a numbered colored plate is stuck on the back of each bee. That's not so easy. The bee must be held in such a way that it can no longer wiggle. A small drop of glue is placed on top of her thorax and the small colorful plate with the number is pressed on it. That doesn't always work. The most annoying thing is when everything sticks successfully, but in the end the bee manages to sting! That hurts even worse: the bee loses its sting and dies. To prevent too many bees from being lost in this way, Gudrun quickly cuts off the tip of each sting with her fine dissecting scissors. Hooray, the bee survives. Only the finger hurts.

Gudrun and Niko help as far as their own programs allow. Three days with sticky fingers. At the end, the 1000 numbered bees are placed in the observation hive. Now Martin has to wait at least 20 days until the marked bees are old enough to start foraging. Only then does the actual and time-consuming part of the work begin. In the meantime, he helps Niko and Gudrun with their experiments. This is how efficiency works, almost like in a bee colony. Martin discusses with Niko at the obligatory kitchen breakfast, where everyone meets every morning, that he cannot carry out his experiments alone. At the same time, he has to log the numbers at the food source and look at the observation hive to see which naïve bees are running after the experienced forager while dancing. Bees fly faster than he can run! Niko remembers that Professor Lindauer bought expensive walkie-talkies for a very similar experimental setup. When Martin sits down at the observation hive with a tape recorder, Niko could use Lindauer's walkie-talkie to transmit by radio which bee is currently at the food source and when it is leaving. Martin records everything on tape. At the same time, he records over the tape microphone which bees are running after a dancing bee in the hive. Niko gives the number of the newcomers and Martin can determine how often the naïve bees have to run after the experienced dancer bees to arrive. No sooner said than done. Professor Lindauer is happy to provide the walkie-talkies. A portable tape recorder with microphone of the brand "Uher Report" is available at the Institute.

The observation hive is located in the Institute's garden. First, an experienced bee must be trained on a food source. Niko takes the bee "Red-57" and places it near the beehive on a food dish with perfumed sugar water. It has to smell like flowers, otherwise it won't work. Red-57 greedily absorbs the sugar water and transports it into the hive.

Soon she will be back at the food dish to get more. If it was worth it the first time, the bee remembers that. Niko has now moved away from the beehive. That doesn't bother Red-57. Again, she absorbs sugar water and remembers the shape and color of the food dish and the scent. Whenever it flies out on a foraging flight, it will first look for this scent and the food dish. She has learned that there is food there. She will not forget that for days. Niko is now gradually moving the feeding place to greater distances. First over the small road, then to the cow pasture and on and on to the fence of "Camp King", an American barracks located below the Institute. Only if the distance is large enough will Red-57 in the hive indicate a clear direction and distance via the dance language. In the meantime, Martin sits in front of the observation hive with microphone and tape recorder and watches the dances of Red-57. He checks whether Red-57 indicates the feeding place correctly and notes which bees are following. Communication via the walkie-talkies works.

Niko: "Red-57 arrived. Camp King. 1:47 p.m. Over."

Martin: "Red-57 dances. Direction correct. Camp King. White-13 follows. Blue-33 too! No, not at all. Loses interest. 1:55 p.m. Over."

They like the "over" at the end of each message. It has something professional about it. It is only surpassed by "over and end" when the experiment is ended. Like in a Bond film. As soon as a marked bee appears at the food source, Niko catches it before it can absorb sugar water. Only one bee should dance at a time, and the followers must remain naïve.

Niko: "White-13 14:05 caught away. Over."

So Red-57 has successfully recruited White-13 to the feeding ground and needed ten minutes to show up at Niko's feeding station. Martin can use the tape protocol to find out how often she had to run after Red-57 to do so.

At the end of the day, they have recorded a number of successful recruitments and Martin has all the important data on his tape. Perfect. This is how it goes on over the next few days. An exciting experiment in which Niko is happy to help. Martin is amazed at the high efficiency of the bees. Often the bees arrive at Niko's feeding station after just one dance was followed. However, the exact evaluation is tedious and will take many weeks. Not all bees are equally smart. Some keep running after Red-57 and never arrive. Martin is not spared from devoting himself to the more than one hundred hours of tape material in detail to see how many smart, medium and stupid bees there are. Niko does not envy him. Everyone has to carry their own dissertation package.

A few days after the end of the experiment, two black BMW limousines drive up to the Institute's parking lot in the morning. Even one black BMW would have attracted attention. Two gentlemen with hats, dark glasses and trench coats get out and ring the bell. They go to the office of the head of the Institute.

"It doesn't look like they're from the beekeepers' association," says Gudrun. She looks up from the microscope under which she is operating on a queen.

Fig. 16. Marked foragers at the feeding dish. In front bee "Red 57" collects sugar syrup.

"Maybe they're from construction. Architects or something. They always dress in black and drive fancy cars."

They are therefore not a little surprised when Agnes leads the two gentlemen to them shortly afterwards. They have taken off their hat, sunglasses and trench coat, but that doesn't necessarily make their appearance any more likeable. Well trained, with parted short hair, but with a tie. Niko assesses them. They move briskly and have a military look, like non-commissioned officers of the Bundeswehr. They are certainly not architects!

"Good day," the taller of the two addresses him: "Who am I talking to here?"

Niko briefly introduces himself and asks back who is standing in front of him. He can also communicate in Bundeswehr mode! The gentlemen then show their service IDs:

"We are from the MAD, Military Counterintelligence Service: Counterintelligence Department. We have the task of talking to you personally."

Where can you find something like that? Does the MAD want to recruit bees as spies?

"If it has to be. I actually have enough to do with my experiments. Does it take a long time?"

They point to Gudrun, who is in the middle of her queen operation.

"We have to talk to you without witnesses."

Gudrun speaks while looking into her microscope: "Unfortunately, I can't interrupt now. I'll be done in a quarter of an hour."

Tall Hendrik shakes his head gruffly: "Is there a room in the Institute where we can talk to you alone?"

"Well, the toilet is too small for the three of us," Niko jokes. "This is a research Institute here. Apart from that, my wife can hear everything I have to say."

The little one takes the floor: "Very well. What doesn't work, doesn't work. If your wife can finish her experiment in half an hour, that would be nice. It is very important that we talk to you alone."

They divide up their behavior, Good cop—bad cop, Niko thinks. Probably fresh from the training course. He then sees the two of them getting into the black BMWs to light a cigarette and wait. Niko is quite dazed by their strange actions.

He is outraged by the arrogance of the two officers: "Should I just drive away? They can look for me wherever they want."

Gudrun knows Niko when he operates on the edge of his control. Now only logic helps:

"You in the VW Beetle against those two in the BMWs? Then I know who wins. Of course, they're stupid, the sooner you get this over, the sooner you'll get rid of them. We are in the middle of the season and we have to avoid all unnecessary difficulties. Our program does not tolerate any complications!"

"What are the complications and what are the unnecessary difficulties?" asks Niko angrily.

"Oh, don't act like that. You have to clarify why the MAD is interested in you and why these agents show up here!"

Gudrun is probably right. The experiments have top priority.

As soon as Gudrun is done with her queen, Niko runs to the parking lot and asks the two MAD officers to come to the lab. As polite as possible. He even offers them chairs. Tall Hendrik pulls a folder out of a black briefcase and places it on the table.

"First of all," says the little one, "we want to give you some background. After that, we will ask you some more specific questions and explain why we need this information."

They ask questions about Niko's military service: Where, for how long, with which units, which courses and which promotions? Aha, so that's the background, let's see what will be the foreground later. They want to check whether Niko knows who he is. He can't see any particular sophistication, especially since the agents know everything anyway, because they compare his information meticulously with their documents. Then there are questions about the Bee Institute and Professor Ruttner:

"Is it true that Professor Ruttner is not a German citizen? And that he often travels abroad? Also in the Eastern Bloc?"

Niko can confirm this: "However, I myself have only recently returned from Israel and do not know any details of Professor Ruttner's travels and appointments in the last ten months."

Tall Hendrik looks at the file: "Yes, we know that. You've been to Israel, eh, on a motorcycle?" He looks at Niko incredulously.

"Yes, of course. How else?"

"Oh, of course. How else?" The two agents exchange glances.

"Yes, that also corresponds to our documents."

The little one now takes the floor: "We would be interested to know when you last entered the so-called GDR, German Democratic Republic?"

"I've never been to the GDR," is Niko's indignant answer. "What am I supposed to do there?"

Tall Hendrik raises his eyebrows: "According to our information, that's not true. You entered the GDR twice on your motorcycle in 1966. Or were you perhaps in West Berlin?"

"Yes, that's right. I visited a family friend with my wife. We have not left the transit highway! It's a awful route on a motorcycle. The ghastly concrete slabs. Terrible controls at the border. Don't you have that in your file?"

The gentlemen look at each other again in silence. The little one changes the subject. "Do you know which military offices of the American army are located in Camp King here in Oberursel?"

Niko shrugs his shoulders and shakes his head: "No, I don't have any reliable information on that. Every now and then, gentlemen in green uniforms walk around. And at the gate there are guards with white helmets. According to what people are saying here, some intelligence department is supposed to be stationed there. Whether that's true or not, I don't know! From the outside, it doesn't look particularly secret. You can see immediately: barracks of our American friends. Why do you ask me? I'm sure you know more about it than I do."

Tall Hendrik wants to flare up. Niko is too snippy for him.

The little one holds him back and answers: "It's quite possible that we know more than you do. However, if we knew more, you would certainly understand that we are unfortunately not authorized to inform you about this."

He smiles, then continues more seriously: "We are more concerned that you have been noticed unusually often at the northern fence of Camp King in recent weeks. You were repeatedly observed there."

Tall Hendrik wordlessly presents him with a black-and-white photo of Niko on his stool at the food bowl, trying to catch a bee. Not a very advantageous shot.

"You were repeatedly at the fence on Sunday, June 11 and the week after. Did you operate a radio?"

Now it dawns on Niko where the rabbit is going: "Yes, I actually used a radio on the fence. A walkie-talkie. I can also present you with the exact times of the radio communication. Martin Dautz has recorded everything in detail! With the tape recorder!"

"You have an accomplice Martin Dautz, with a tape recorder?" The otherwise gloomy eyes of the Tall One flash.

"No accomplice. A fellow student. Martin is also a Ph.D. student. He wants to find out how many bees a dancing bee can recruit to a good nectar source."

"Dancing bees? What's the point? Recruits?"

The little one apparently finds it difficult to classify the facts. Niko is now in his element and explains to the two in detail how the experiment is set up. However, he quickly realizes that these gentlemen from the MAD not only lack any basic knowledge of bee biology, but that they also do not understand the logic of the experimental approach. In the laboratory hangs a board that explains the bee dance graphically. Niko usually uses it to teach his students. Now he can use it to explain the secrets of honey bee communication to the two of them step by step. Their questions show that Niko's didactic skills are not enough in this difficult case. They usually miss the point. Finally, he limits himself to the current processes and the description of radio traffic.

"So: I announce which bee is currently landing at the feeding site, and Martin records it on tape."

The MAD officers are sitting more relaxed in their chairs. Niko does not believe that they have even come close to understanding what it is about. They seem to realize that you couldn't come up with such a crazy excuse if you were a spy.

The little one even offers help: "We also have all the radio traffic. We got everything from the Americans and made a copy. The US experts could not decipher the code and consulted us."

Now everyone laughs together. "Before we leave, one last question: What does Red-57 stand for? This has caused particular concern among the Americans. You know, the enemy is red."

Niko is now sure that they have not understood anything of his explanations and stands up.

"I'll call Martin. If we're lucky, we can show you Red-57 in person right away."

Martin is very insecure at first. The MAD is interested in his experiments? Is he in the sights of the MAD? Niko calms him down. They go into the bee garden and Martin carefully opens the wooden doors of the observation hive so that they can now see the bees through glass panes.

"There! A red one!" Tall Hendrik has seen something.

"Yes, exactly, red-25," says Martin. "There's yellow-57. Also wrong".

At the same moment, the bee with the red tile No. 57 lands at the entrance hole and runs into the colony.

"There's Red-57!" Martin is very excited. He now knows her personally, as a very reliable collector. It recruited by far the most followers for him. On the comb, it passes on its food to another bee. She dances briskly in the direction of Camp King, only now a longer distance, as Martin easily realizes. In the middle of Camp King? He prefers not to pass on this suspicion to the agents. The two agents are satisfied and happy to have solved their difficult case so confidently. They apologize kindly to Gudrun for "the intrusion".

"After all, we have to do our job! Regulations are regulations!"

"Brain is brain," Gudrun thinks and says friendly: "Have a nice day."

The German "Aufwiedersehen" ("see you again")? No, she certainly doesn't want to meet them again.

As quickly as the black BMWs had come, they are gone again.

Gudrun and Niko look silently at the parking lot for a while, where their Beetle is now standing alone again next to Professor Ruttner's Variant.

"It can happen so quickly. You can be glad that they didn't take you away and hook you up to the lie detector right away. They are apparently allowed to do everything, anywhere, at any time."

"In any case, they know everything, from everyone, from anywhere, at any time, and that's scary enough in itself."

Nineteenth chapter

Gudrun and Niko make a far-reaching decision

Shortly before the end of the season, everyone at the Bee Institute is trying to review important parts of their respective work programs. Most of the time, the bees—now preparing for winter—no longer play along. Martin's observation hive with many marked bees is robbed by colonies from the apiary. Professor Ruttner's experiments to raise new queens more or less fail. Only a few queen cells are accepted by the foster colonies but the hatched queens are too small for instrumental insemination. Gudrun's operations on the queens continue to go well. Many of the operated queens do not continue laying eggs. Niko's experiment to induce queens to lay fertilized eggs on drone combs by transporting the experimental colonies to the still flowering Gamander about 100km south also fails. An extended period of bad weather finally convinces even the most stubborn among them: That year's bee season is over. Professor Ruttner aptly puts it: "After this season comes the next season and then we have to do everything which we didn't manage to do this time."

For Gudrun and Niko, a review of their results is on the agenda. They arrange protocols and evaluate results as good as possible. The first summary gives a clear picture: "The 1967 season was successful for both of them overall."

Almost all queens have not only survived up to 90 days, they have continued to lay eggs and the young queens have even gone on their mating flight. Next summer, Gudrun only has to confirm her results with about 20 queens: The tracheal network is important for the survival of the sperm, the gland is crucial both for the migration of the spermatozoa after the mating flight and for the subsequent fertilization of the eggs. She has to write this down in her dissertation. Gudrun, who tends to be critical, cannot be happy about her great progress.

Niko says: "In my estimation, you should be more than proud and very satisfied. In the last twelve months, working in Israel, you have exceeded even the most optimistic expectations. Stop whining and rejoice!"

"Niko, you are incorrigible. Your glass is always half full! When I think about what is still missing, my glass is really half empty."

"Gudrun, did you listen to what Professor Ruttner said after reviewing your summaries? I repeat his comment: "It is foreseeable and very likely that you, Gudrun, will be able to complete all the experiments necessary for a dissertation next season." Besides, I'm hungry: Let's go to Sachsenhausen for ribs and cabbage!"

"Wait, I'm just getting my things. I, too, am hungry as a bee after too long a break from foraging." And Gudrun laughs.

It was not only her experiments that went well and showed promise this summer. Many well-known scientists visited the Institute. Karl von Frisch, W.C. Rothenbuhler, Anna Maurizio, C.G. Butler, J. Woyke and Harry Laidlaw. Professor Ruttner introduces them to his students and proudly reports on their results. Professor Jurek Woyke is particularly interested in the operation on the queen bees: "Is it really true that the operated queens fly out to mate? I can't imagine that. Am I allowed to watch an operation?"

Gudrun agrees. After the queen has recovered from the anesthesia, he wants to take a close look at her. Professor Woyke, who had held many thousands of queens in his hand, takes the operated specimen—and suddenly the operated queen escapes from him and flies across the room. Near the large window, it gets caught high up in a spider's web. Gudrun is horrified and speechless. Niko shows presence of mind, quickly fetches a ladder and frees the escaped queen.

"That has never happened to me before!" says Professor Woyke apologetically.

Gudrun has the feeling that he wanted to check whether operated queens can still fly. Without a word, she frees the queen from the sticky cobwebs and carefully returns her to her colony. What luck: This queen survives and goes on a mating flight as expected.

At the beginning of the winter semester, Gudrun and Niko are the only two in the Biological Institutes. As assistants, they supervise various internships. At the same time, they complete further courses in order to obtain the last necessary certificates for admission to the Rigorosum. Next fall, they would probably have to meet all the requirements for the doctoral examination.

In November, they hear that the extension of the Bee Institute has been completed. Professor Ruttner proudly leads them through the new premises. He has planned the space allocation. Gudrun and Niko are given the first, very large laboratory on the ground floor.

"Don't make such a face, Niko. Rejoice and say thank you for the beautiful new laboratory. We finally have enough space here, and a flue for my chromatography and chemistry."

"Gudrun, you're heartless: Our little room in the old building, where the legendary

Hugo Gontarski did his research, has a special academic atmosphere. Creativity and narrowness, a combination that we have now lost? That makes me sad!"

"I don't give a damn about your Spitzweg idyll! What counts for us are the practical improvements in working conditions, not unrealistic sensitivities."

"Unfortunately, you're right in principle, because if we continue to pursue a career in research, it won't be possible in the old witch's house like this. The Americans, who have offered us scholarships during their visits, were very surprised about the working conditions here."

For Niko, the future together is clearly and obviously firmly planned. He wants to continue researching honey bees together with Gudrun. The whole world is open to them. Bees and honey have a high international status. And after completing their doctorate, both have promising prerequisites for scientific careers!

On the other hand, Gudrun begins to have doubts as to whether science alone will be enough for her to have a fulfilled life. Due to the strict hierarchy among the professors and the competition among assistants and students, the freedom she dreamed of was clearly restricted. She has been thinking about other plans for some time. Ultimately, the desire to have children of their own comes more and more to the fore. The discrepancy between her desire to have children and Niko's plans increasingly affects her mood and leads to a reduction in her productivity. What happens if Niko rejects her desire to have children of her own? Research projects worldwide, independence and unlimited flexibility could be more important to Niko than children and family? Then Gudrun remembers babysitting together when studying in Freiburg. Yes, these were harmonious and positive experiences. Obviously, not only she but also Niko enjoyed taking care of the children at that time. In the meantime, however, the doctoral thesis and the entry into university teaching have largely dominated Niko's activities. He also didn't notice Gudrun's problems with a scientifically oriented future perspective. He has blocked her previous, probably too timid attempts to rediscuss the plans for the time after her doctorate several times. Gudrun realizes that she urgently needs to talk to Niko. Whether a visit to Freiburg would not offer a good opportunity to discuss her plans in detail with Niko? And indeed, her suggestion to go to Freiburg together next weekend and visit old friends is enthusiastically welcomed by Niko.

"Lunch at the restaurant "Schönberger Hof" high up on the mountain! I'm dreaming of Schäufele with Kraut!"

They sit in their Beetle early on Saturday and drive comfortably south on the motorway. After Karlsruhe, the weather clears up and shortly before Freiburg a blue sky and the sun greet them.

"Niko, let's park the car near your old student's room. There is still time enough. From there we will reach the Schönberger Hof in time for lunch," suggests Gudrun.

The path to the Schönberg first goes past the Jesuit Castle through meadows and fields. They walk slowly and talk about shared memories.

Gudrun swallows and continues: "It's no coincidence that we're here! Let's find a place to sit down. Niko, I have to talk to you!"

Niko looks around searchingly and then walks up the steep slope for a while. He has found a felled beech. There Niko spreads out his anorak:

"Come Gudrun, we can sit here. Tell me, what's going on? Did I do something wrong?"

Fig. 17. The extension of the Bee Institute in 1967.

Gudrun slowly climbs up the slope and sits down next to Niko.

"No, it's not about you, it's about me first. At the age of 24, it's time for me to think about children. Our own children!" Gudrun blurts out and looks at Niko expectantly.

Niko returns her gaze and looks into Gudrun's green eyes:

"Dear Gudrun, you are always good for a surprise. I remember our discussions during the first semesters. Gudrun at the time did not want a family and no children. Just be free and go full steam ahead into science. That was and still is my thing! A good, successful and productive basis for our common path. And now you're talking about children! "

"I was 19 or 20 at that time. I came from a family with many constraints and duties. Wilhelmshaven and the school were also characterized by an oppressive crampedness! The world of science, that was my path to freedom, in short my dream!" Here Gudrun takes a break.

"And this dream is now over?" asks Niko.

"No, not at all! However, I have recently become somewhat disillusioned. Yes, bee research and science are great. The strict hierarchy, coupled with the extensive dependence of subordinates and junior staff, as well as the fierce competition between scientists, often even between students, these are major deficits! For me, empathy and social interaction are important and these are rather rare exceptions in science. Yes, I want to and I will complete my doctorate, but after that to continue to focus on bee research and science, I don't want that! A life without my own children is not desirable for me. I should have talked to you about it earlier. You were completely eaten up by your doctorate and university, so I didn't get through to you. On the other hand, it was perhaps also the fear of this principled discussion that prevented me from defending my wishes more vehemently."

Gudrun doesn't talk any further. She is aroused and snuffs into her handkerchief. Niko is speechless. He puts his arm around Gudrun and pulls her to him.

"Stop it, you're crushing me!" protests Gudrun.

Niko senses that this protest is not meant seriously and continues to hold Gudrun in his arms—not quite so tightly.

"Niko, I need your answer! Finally say something!" demands Gudrun.

"You know my answer. I want to be with you. You can count on me. Anytime and whatever! I will then be the bee researcher at your side and look forward to our children!"

Gudrun presses the handkerchief to her eyes. Niko sees a tear running down her face and falling into the grass. What can he do?

"Gudrun, I'm hungry. If we don't start immediately, we'll miss our Schäufele at the Schönberger restaurant."

Gudrun frees herself from Niko's arms and stands up.

"Slowly I have the feeling, Niko, you can read my thoughts and feelings, but it didn't work out with my desire for children!" laughs Gudrun happily and runs off.

In the early afternoon, they arrive at the Brinkmann family's house, who has offered them a place to stay for the night. Herbert also brings the usual brewed coffee. Obviously no 'changes'. As then, the children dominate the entertainment. Suse and Julius are at school, so there is a lot to talk about. Caroline has just started kindergarten. Although she hasn't seen Gudrun and Niko often, she lets herself be infected by her siblings and desperately wants to play with them. Maria is pregnant again. Everyone protests when Gudrun and Niko drive to Freiburg in the early evening to meet up with their old friends from their first semesters.

"We'll definitely be back in time to say goodnight to you."

"And what about the bedtime story?" the "big ones" grumble.

"Maybe it will be enough for a short story," Niko comforts.

Of course, Michael, Uff and Werner are waiting in the Wolf's den. Gudrun and Niko are greeted with a big hello. After the thick Schäufele at noon, they order only a "Strammer Max (Hamburger with ham and eggs), and Niko a Pinot Noir. Gudrun drinks water. She has promised to take over the return journey in the car. The conversation revolves mainly around their research work. The three Freiburger friends can't believe that Gudrun and Niko expect to complete their dissertation next winter.

"That's not possible at all. We are still in the middle of our experiments for the diploma. Only with a lot of luck can we get our diploma next year and only then start with the doctoral thesis," explains Michael.

The "Frankfurters", on the other hand, cannot imagine that. "We were lucky! The diploma for biologists will not be introduced in Frankfurt until next year. Gudrun and I are, so to speak, old cases and are allowed to do a doctorate without diploma," laughs Niko.

"The experiments for a diploma thesis are limited to a maximum of 6 months, as far as I have heard. You've been working on your work for over a year now, haven't you? How does that work?" asks Gudrun.

"Yes, officially only half a year is allocated for a diploma thesis, but the professors can apply for an extension. And that's what everyone does here! Normally, a diploma usually takes 2 years or even longer due to these extensions!"

"A typical exploitation of students! High-quality experimental results without paying appropriately. When it comes to exploiting exam candidates, professors have no inhibitions," Gudrun enthuses.

"That's right, Gudrun. I am mainly interested in creating a "very good" grade and so I will not say "no" if my supervisor wants an extension. The more time I have, the better the work gets!" Uff notes.

It goes back and forth for a while before Gudrun and Niko leave for Bollschweil at around 8 p.m.

There they are eagerly awaited. After a short story and an extensive goodnight with the children, the adults finally sit together with a glass of red wine as they used to.

"Yes, you're right. With three children now, this apartment is really much too small!" Maria answers a corresponding question from Gudrun. "Before the new child comes, we want to move. It's all complicated because of Herbert," Maria complains.

Herbert is irritated: "Of course it's not my fault! However, Maria probably means my application to the Free University in West Berlin. I might get a call to be there soon. And Maria wants to wait with the search for a larger apartment until that is decided. I, on the other hand, would rather look for a new apartment in Freiburg today than tomorrow. We can then wait in peace until the birth is over. If the professorship works out, there is still time to move to Berlin," Herbert notes.

"Typical Herbert, better two moves instead of one! For him, none of this is a problem. The work and the furnishings are primarily up to me," Maria complains.

"Herbert, if I understood you correctly, are you in first place on the list?" asks Niko. "Gudrun and I congratulate you! A professorship at the Free University in West Berlin. That's a great honor!"

Niko raises his glass: "We want to toast your professorship in Berlin. Everything, all the best and good luck, cheers Herbert, cheers Maria!"

Then a question from Maria, for which Gudrun had been waiting: "Gudrun, what's next for you?" With a quick finger pointing to her belly, which is big because of the advanced pregnancy, she continues: "Are you planning to finally have children? How many years have you been together now and always only chaste science?"

Niko is quite shocked by this direct statement by Maria. He wants to say something about "personal remarks" that are not welcome. Gudrun is faster:

"Dear Mary, our relationship is less chaste than you imagine, despite or perhaps because of the research! You're right, it's time to think about children. Just today at noon, Niko and I talked about it. I'm 24 and so we don't want to wait any longer!" beams Gudrun.

Niko is rather embarrassed by this turn of the discussion. He is thinking about how he can bring the issue to a close as quickly as possible. Herbert intervenes: "Gudrun, don't you want to wait until you've finished your doctorate? If I remember correctly, there are concrete prospects so you will finish next winter?"

"Herbert, good, beautiful and very reasonable," Gudrun answers vehemently: " I don't want to plan, as I did with my experiments. I want a child and as soon as possible. I have subordinated everything to bee research long enough. I don't want a life without children!"

Maria supports Gudrun: "If you think long enough and, above all, without emotion, there are always 1000 reasonable reasons why you have to postpone a desire to have children. We are liiving in a very cramped apartment and Herbert is waiting for his call to Berlin. There is probably no more inopportune time for a fourth child. I'm looking forward to this pregnancy and even more to the baby! Gudrun, doctorate or not, your desire to have children comes first!"

Niko is uncomfortable with this discussion.

"Dear people, we have to leave tomorrow morning and your old guest sofa is calling! Thank you very much for the nice evening! All the best and let us hear from you with " Child Number Four" and the call to Berlin!" Niko gets up and so does everyone else.

With Gudrun's pregnancy a little later, "together" facts are created. Gudrun now "only" has to finish her dissertation. And after that, Gudrun is pleased, only motherly happiness and housewife. She continues, relieved, and at the same time a little worried:

"That looks different to you, dear Niko! You have to pass the exam and look for a job as a scientist. For me, the competitive pressure is over with the dissertation, for you, Niko, it really starts afterwards!"

Accordingly, Gudrun worries little about her exams and experiments. She happily tells her parents on the phone that she is expecting a child.

"That's what we've been afraid all along, that you won't finish your studies!" Gudrun's father is outraged.

"Of course, I'll finish my doctorate before the birth. Definitely," Gudrun assures.

"You're only in your twelfth semester! You'll never make it!"

Her mother also expresses similar concerns very directly. Gudrun can't believe it. At Niko she cries herself out:

"What a pity! Not a bit of joy for the future grandparents. And no understanding at all for the fact that I am happy!"

Full of fear, she now goes to Professor Ruttner to inform him of her pregnancy. This one beams at her:

"My heartfelt congratulations! Gudrun, are you happy?"

"Am I happy? Yes, of course, I am looking forward to our child with all my heart," is her answer.

Twentyieth chapter

Gudrun and Niko furnish a new apartment and submit their doctoral thesis

While Gudrun continues her experiments in a good mood, Niko is worried about finding a suitable apartment. With a baby in her arms, Gudrun can hardly climb up and down the narrow ladder to the "barrel". And there is only a small sink in the unheated anteroom. No bed, no matter how small, fits in the room. They talk about these problems with the landlady. Mrs. Werner congratulates warmly. She is visibly relieved that

Gudrun and Niko want to look for another, larger apartment for their baby. Yes, Mrs. Werner wants to ask around and let them know when she finds an apartment.

Niko registers with some real estate agents, including Oberursel. This costs an initial fee, but all the offers they receive are beyond their financial possibilities. The small 3-room apartment in Bockenheim that Mrs. Werner found also costs more than 300 marks a month, five times as much as her current room. Niko despairs.

"It's still a long time until the birth in December. More than six months," Gudrun tries to reassure.

When all the queens in a series die of an infection a few days after the operation in the summer, Gudrun's optimism gives way to a certain nervousness. Professor Ruttner helps and gives her queens as replacements, which he had actually planned for his own research. After careful disinfection of the equipment, all queens now survive. And the results achieved are in line with those of the previous year. However, this delay puts Gudrun's schedule in jeopardy. Everything has to happen quickly now. The time of birth cannot be postponed!

When they still haven't found an apartment after a few weeks, they finally try the student accommodation service at the university. The employee currently has no apartment, in general they only have offers for rooms. He offers to put Gudrun and Niko on a waiting list. Niko can no longer control himself. After fierce insults, he declares loudly:

"A baby is on the way and needs a place very soon. What an outrage with your waiting list. Our child will not wait."

An older woman approaches them and kindly invites Niko and Gudrun into her neighboring office. Gudrun leads the way and so Niko has no choice but to follow. Before he can rant any further, Gudrun says clearly:

"Niko, calm down. Your bullying doesn't help us and our baby."

The employee explains to them in a friendly manner that there are apartment addresses at the accommodation service, but only for social hardship cases:

"You could have explained your situation right away and calmly!"

Niko apologizes: "I'm very sorry. we can only pay a monthly rent of up to 200 marks and have been looking in vain for weeks now."

They receive the address of a landlord in Bad Homburg, and leave the student accommodation service with ease.

For Gudrun and Niko, Bad Homburg is perfect. From there, the way to the Bee Institute is shorter than from Frankfurt-Bockenheim. The owner of the apartment, Mr. Euring, a white-haired man, greets them warmly. In the living room they also meet Mrs. Euring. Mr. Euring wonders where they got their address from. His wife probably informed the student accommodation service without her husband's knowledge.

The Institute for Beekeeping in Oberursel is well known, and the Eurings seem to understand that they are now looking for a larger apartment after the small room in Bockenheim. Then their concerns come up: "You have to know, this is an old apartment on the first floor with stove heating and no bathroom. Are you sure that would meet your requirements?"

Gudrun and Niko study the floor plan: a living room, a bedroom and a kitchen with a pantry, the toilet on the floor, but in front of the apartment. However, not one ladder as in Seestraße. "I grew up with coal stoves, I still know how to heat with them," explains Niko.

"The rent is 170 Deutschmarks plus water and electricity. Unfortunately, a viewing of the apartment is not possible today." In the course of the week, they could make an appointment for a viewing.

While drinking coffee at the sparkling clean kitchen table, Niko suddenly explains: "We want to forego the viewing appointment and rent the offered apartment now. We have been looking for an affordable apartment for too long." He puts a DM 50 note on the table as a deposit. Gudrun is surprised by Niko's sudden decision. After the long search for an apartment in vain, she agrees not to let this opportunity pass by unused. This apartment offers enough space for the baby. Mr. Euring does not want to accept the deposit at first. He thinks that even without a down payment, his promise of the apartment should be sufficient. Niko then explains that he can only sleep well again when they have paid for the apartment and then adds:

"My wife can testify to the many sleepless nights."

Mr. Euring laughs. Mrs. Euring asks her husband to take the money. Does she suspect that Gudrun is expecting a child? They didn't mention pregnancy. In the end, Mr. Euring takes the money and gives them a receipt. Mrs. Euring calls her son, with whom they arrange a viewing of the apartment next week. "Then you come back to us and sign the lease."

Gudrun and Niko thank them warmly: "We are so relieved and will finally sleep well again. We are very happy about the new apartment."

Her next visit to Eurings leads to a clarification of the situation. Mr. Euring had advertised the apartment in the newspaper one day before Gudrun's and Niko's visit. Eurings have received more than 20 requests from this advertisement. Among them is a young couple who are expecting a child soon.

"Yes, if I hadn't taken the deposit from you, this couple would have gotten our apartment!" says Mr. Euring.

Gudrun and Niko can comfort him: "We also have a baby on the way! That's why we were so desperate!"

Mr. Euring fetches a bottle of wine. They all toast to the coming baby! Only water is poured into Gudrun's glass. Together with Eurings, they are pleased that it turned out as desired by everyone.

Gudrun notices that Niko now observes her waist circumference more often with satisfaction. "Yes, it'll be okay," she thinks confidently. She has no doubt that Niko will lovingly take care of their child together with her.

The move from Bockenheim to Bad Homburg does not cause any difficulties. They mainly own two typewriters and some books. The attics of friends are rummaged through, and what is still useful finds its way into the new apartment. Also, a cot, and even a nice big aquarium. Baby clothes are promised by Gudrun's sister Ute, who has just given birth to her third healthy son. At the birth of Gudrun's child, this baby equipment will be available. Everything in blue, it doesn't matter! Two investments are made from the scarce funds: Gudrun gets a good bed with a first-class mattress, and a new refrigerator is set up in the kitchen. Gudrun's parents donated the stroller, Niko's parents a washing machine. Aunt Anna finds a frame with wheels for a bassinet and the accompanying bucket, which she used as a laundry bucket for a while. She lovingly prepares it again into a baby bucket, with fabric ruffles and even with a "canopy roof", all in yellow. Gudrun doesn't know whether it will be a boy or a girl. She doesn't care, the main thing is a healthy child! She is very grateful for this help. She has—as she says—no time for "such things".

"I want to have time for the baby when it's here. That's the most important thing! That's why I have to evaluate the last experiments before the birth, write down the work and pass the exam. I can't take care of the equipment of the children's room at the same time. I'm very much looking forward to our child."

"Gudrun, you're right as always. Daughter or son is also welcome to me. You will certainly manage everything well and be a loving mother who has a lot of time for our baby," jokes Niko.

In this situation, Gudrun's mother is a great help, she types the first copy of the doctoral thesis with five carbon copies for both of them on her old typewriter. She is not the only one to help. Gudrun's pregnancy shows clearly and it is mainly women who help with the completion of the illustrations for the written documentation. Slides and drawings are preferably completed for her at the Zoological Institute. The administrative staff organize a quick run-through of the doctoral thesis. They make sure that the work does not remain with a professor for more than two days. Gudrun is happy about this solidarity. She would not have been able to speed the procedures by herself.

Niko's work, on the other hand, remains with the reviewers for a long time, until Gudrun finally desperately asks the secretary for help for the "father of my child:" "We both have to finish exams before the birth. Especially because my husband has to find a job quickly." And contrary to expectations: Now the accelerated completion of the dissertation also works for Niko.

Twenty-first chapter

Gudrun and Niko complete their doctorates, become parents and do not accept a lucrative job offer

In the oral doctoral examination, the Rigorosum, basic knowledge of biochemistry, botany and zoology is tested. Gudrun and Niko have to master the standard works in these disciplines—an almost impossible undertaking! Immediately after the submission of the written work at the beginning of November, there is only cramming. Gudrun has difficulty concentrating because her baby is kicking in her belly.

At the end of November, their theses on the dissertations are accepted and the oral examinations are scheduled for December 10 and 12. The gynecologist has calculated December 27 as the due date!

"That's very close!" remarks Niko's mother, an experienced pediatrician and mother of 5, laconically on the phone. "Gudrun and you Niko, you've always been lucky so far. That's why your baby will wait and only come after Gudrun has passed the exam. A mother with a doctorate is not to be sneezed at, as you, dear Niko, probably know best yourself."

Gudrun's zoology exam with Professors Lindauer and Professor Ruttner goes well. She can't answer Professor Ruttner's question about how much oxygen is bound in a liter of blood. Gudrun starts with the structural formula of hemoglobin, which she loosely draws on the pad in front of her, including the oxygen molecules. She knows the amount of hemoglobin in the blood and calculates the correct value for oxygen binding based on general metabolic data. Instead of criticizing her "lack of knowledge", both examiners are impressed. Lindauer is very pleased with her calculation:

"This shows creativity and solid scientific understanding."

The examiners do not know that Gudrun had synthesized hemoglobin during her chemical internship in Freiburg. Gudrun waits contentedly for Niko, who is next to her. He also comes out very confident:

"Gudrun, everything is fine? It went well for me too."

The exams in the other subjects are also going well. Of course, this is celebrated. First, Professor Ruttner invites Gudrun and Niko to the restaurant in the castle of Bad Homburg for a flambéed entrecôte double. He is very proud of his first students, who have achieved such good results and are recognized by his colleagues.

Less distinguished things continue with the friends. They go to Freiburg and invite their old friends to the Wolf's Den. It just has to be! Gudrun likes it wonderfully, whether posh or hearty, and, it seems, so does the baby in her belly. In any case, it will remain there for a little while and will weigh 3750 grams at birth. Even Gudrun's parents are finally happy and extremely satisfied that their daughter has "against expectations" achieved her doctorate.

Niko's parents were not worried in this regard: "After all, our son comes from a family with a long academic tradition."

And then finally comes the big day. On December 30, 1968, three days later than calculated. Gudrun is sitting comfortably in the armchair in the evening when suddenly everything around her is wet. That must have something to do with birth. Niko quickly fetches her packed suitcase and wants to go, but she wants to put on something dry first: it's snowing and freezing outside. Niko is nervous and urges her to hurry. His driving is impossible, much too fast for black ice and snow.

Gudrun gets angry: "Niko, if you have an accident now, we'll never get to the hospital in time. Please drive carefully! Otherwise, I'll get out!"

A little slower now, they reach the hospital. Niko is still allowed to help at the reception, but in the district station entry for men is generally prohibited. So, he has no choice but to go home. Every two hours, he walks through the falling snow to the phone booth and calls the hospital. In the early morning, he falls asleep exhausted. Before 7:00 a.m., he drives to the Institute, where there is a telephone. The cleaning lady is there and congratulates him on his child. The hospital called there on behalf of Gudrun. He immediately jumps back into the car and races to the hospital. There is no Gudrun next to him to slow him down.

Gudrun lies beaming in bed with her baby: "Niko, it's a girl! Here, take her in your arms!"

Niko hugs them both. When he picks up the baby, it wakes up and looks at him contentedly.

"We should call her Anna Franziska! What do you mean? As a thank you to our Frankfurt aunt Anna and to my mother, both have helped us a lot, especially in the last few weeks. My grandmother, whom I cared for at the time, was also called Franziska," Gudrun suggests and looks at Niko expectantly. He agrees: "I'm quite sure that Aunt Anna will feel honored, even if she may not admit it."

And he's right, because the aunt's comment is: "The name Anna is really from another time. At that time, it was a common name for maids!"

Gudrun's mother, on the other hand, shows her joy: "How beautiful! Then the name Franziska will go into the third generation."

When Gudrun is allowed to go home after a week, Niko has prepared the apartment well. Everything is clean and the coal stoves spread a cozy warmth despite the cold outside. The baby basket with canopy and wheels proves to be the ideal solution. The microclimate remains almost constant in this small, enclosed room and Anna can always be pulled along. Niko can still stay at home; the bees sit quietly in their winter cluster and his scholarship has not yet been approved.

He takes care of Anna almost all the time, bathes and changes her diapers. Gudrun starts to complain: "It's a good thing you can't breastfeed Anna. Otherwise, I wouldn't be able to cuddle her at all."

Gudrun and Niko had hoped to finally become financially independent of their parents after their successful doctorates and to earn their own living. The scholarship Niko applied for from the German Research Foundation has not yet been approved. They are not spared from asking their parents for further support. The answers are positive and understanding. Niko's and Gudrun's parents assure that they will support them as long as Gudrun and Niko want. Niko's parents even offer to top up the monthly payment by an amount for Anna: "There are three of you now!" Niko rejects this in a friendly but firm manner.

In this situation, Helmut, a friend and former colleague of the zoological internship, comes to visit. Helmut completed his doctorate in Frankfurt a few months earlier and now works at BASF in Ludwigshafen. He invites Gudrun and Niko and Anna to a good dinner at an Italian restaurant and talks about his work: Helmut teaches biology to laboratory technicians and technical staff. A job that he enjoys and is paid "decently".

"BASF is looking for more scientists to train employees. We worked well together on the supervision in the zoological internship. Maybe you would like to apply?"

This suggestion is probably the reason for his visit.

"Together with you, Niko, we could redesign biological training at BASF."

Helmut has seen the "elementary living conditions", as he calls them, of Gudrun and Niko. "You are living almost in poverty at the moment. The modest starting salary at BASF is a minimum of DM 2500. Without a doubt, a significant improvement in your economic situation and quality of life."

At first, Niko and Gudrun are very pleased with the offer. They thank Helmut for the visit, the good food, and for his trust.

"Niko has applied for a research scholarship. Your tempting offer surprised us. We need time to think. He will call you in the next few days," says Gudrun at Helmut's farewell.

It soon becomes clear to them that with such a job, the dream of life as a bee researcher would be over. Gudrun is primarily affected by her "cramped" economic circumstances as a mother and housewife.

"I want to follow your decision," Niko says later.

He drives to the Institute in Oberursel and when he returns, he sits down in front of the aquarium without a word.

"You are silent, Niko? Finally say something!" Gudrun is annoyed.

"What can I say?", Niko asks. Then he says: "Gudrun, you're right as always. Yes, we have to consider whether this opportunity at BASF is right for us. I have to sleep on it first. We will decide in the course of the next week."

For a few days, they do not talk about BASF. Their thoughts constantly revolve around this decision. A scholarship of 1,000 marks from the DFG, which has been applied for but not yet approved, or a permanent position in one of the most renowned German companies with a very adequate salary? Gudrun finds the idea that Niko should give biology lessons for technical assistants from now on reasonable and terrible at the same time. Niko does not comment.

"Niko, whether you like it or not, we have to talk tonight!"

She is determined to finally come to a decision. She puts daughter Anna to bed, then fetches a bottle of red wine. They toast.

"Niko, you have to wait until the DFG approves your proposal. Employment in industry is possible at any time. Now continue your path as a bee researcher. Bee research was and is also important to me!"

Niko no longer thinks: "Yes, Germany is still in the "Wirtschaftswunder" (economic miracle) mode, that will certainly continue for some time. There are many vacancies in industry, so no problem if it doesn't work out at university later on. That's our luck. We are the post-war generation and we are needed."

Nevertheless, he has to address the good salary at BASF. Gudrun laughs:

"Better a happy bee researcher with bulky waste in his apartment than a frustrated industrial biologist with designer furniture."

Niko is relieved: "Dear Gudrun, no question, we can continue to be happy with bulky waste in the future. Maybe we could have new furniture and bee research at the same time? Let's at least try that!"

When Niko calls Helmut and cancels, Helmut is surprised and disappointed. Before he hangs up the phone, he sends his warmest greetings to Gudrun. Obviously, he had the impression that Niko had made this decision without Gudrun's consent.

Twenty-second chapter

Gudrun and Niko stand on their own two feet financially and participate in the World Bee Congress

The approval of the research scholarship of the German Research Foundation will be delayed until March. Finally, Gudrun and Niko are financially on their own two feet.

"We have finally achieved this important goal: We are no longer dependent on our parents' wallets. Fully grown-up," Gudrun says happily.

"Although our career goal did not meet their wishes, our parents supported us for six long years and six months. We owe them a great gratitude!" Niko sounds a bit pathetic.

"Perhaps we should pass on this generous support to our own children in due course, even if they – God forbid—decide to become bee researchers!", Gudrun ends the speech.

The research scholarship comes in good time before the start of the bee season. Not much has changed for Gudrun and Niko. Daughter Anna is packed up in the morning and comes with her to the Bee Institute. The large laboratory offers enough space for the pram and then also for a "stable". At the age of a few months, Anna still needs a lot of sleep, so that Gudrun and Niko can continue their experiments. Gudrun can perform further operations on queens. Professor Ruttner has worked out a corresponding plan for her: Cutting the glandular branches without removing them: "Do the queens lay fertilized eggs afterwards? What happens after a puncture of the spermatheca with intact glands?" Above all, the experiments on sperm chemotaxis are fascinating. Can sperm distinguish between extracts from glands and spermatheca?

He entrusts Niko with the preparation for the major internship, which is to take place in Oberursel for three weeks in the summer semester. This requires a convincing and attractive concept. Experiments for 16 course participants have to be planned and prepared. On the one hand, the program is intended to provide an introduction to the social life of bees, and on the other hand, the internship must inspire the participants for bee research.

"In this course, we were supposed to win the best students for a thesis with us!" says Professor Ruttner during one of the many preparatory talks.

For this summer of 1969, there are two more commitments. Gudrun and Niko are to help with the organization in August of Apimondia, the World Bee Congress, in Munich and present their doctoral theses in lectures. Gudrun's eldest sister Ingrid is willing to take care of Anna for two weeks, who will be eight months old at that time. At that Congress, Gudrun and Niko are successful with their lectures. And at both, they are offered scholarships in the USA. Professor Ruttner complains to his colleagues that they made the offers without asking him first: "Niko's scholarship with me is still running, and then we can apply for a new one!"

At the Congress, they find a variety of suggestions and hear about new scientific results that captivate Gudrun and Niko. Niko is particularly fascinated by a lecture by Professor Roger Morse: It is about the biology of giant honey bees in the tropical rainforest of the Philippines. Only the special suits of the Montana firefighters offer sufficient protection against the long stings of these large, defensive bees. Several times Gudrun feels Niko's elbow in her ribs and she hears him whisper:

"I have to work with those bees!"

Following the Congress, many well-known bee scientists visited Oberursel. Roger Morse has a scholarship to offer for a "postdoc" and invites them to Cornell University in Ithaca. Unfortunately, the project there is not about giant honey bees, but about the mating of queen bees in cages. This project does not seem very promising to Niko and Gudrun. Finally, Professor Ruttner intervenes and makes it clear to Roger Morse that he needs Niko at the Bee Institute and wants to keep him in Oberursel. Instead, he negotiates an offer for Martin Dautz, who now also has a doctorate. So instead of Niko, Martin goes to Roger Morse in the USA for a year. Nevertheless, Morse hopes, he says, to be able to welcome Gudrun and Niko to Ithaca sooner or later.

After the bee season, daughter Anna energetically begins to explore her environment. Gudrun now stays at home more often and watches her progress with great pleasure. Once Anna has achieved what she wants, she looks at Gudrun beaming. She waits for her mother's praise, which Gudrun gives abundantly. Gudrun is also working on the publication of her doctoral thesis and is now more or less cut off from the Institute. She is all the more pleased about visitors from the Institute, who often visit her in Bad Homburg.

Twenty-third chapter
Niko fetches bee colonies in Pakistan

Since the summer, Professor Ruttner has often discussed his own research projects with Niko. One focus of his program is on studies of the eastern honey bee, *Apis cerana*. Its natural distribution area is in Asia, geographically separated from the western honey bee, the *A. mellifera*. It is similar to its western sister species in numerous biological traits and characteristics. The similarities go so far that, for example, the internationally important ancestor of bee systematists, Professor Dr. Hugo Berthold von Buttel-Reepen, considered the eastern honey bee only as a subspecies or race of the western one, i.e. as *A. mellifera cerana*. Other researchers, including Professor Lindauer, who studied the bee dances of Asian honey bees in Sri Lanka, are of the opinion that the eastern honey bee is a separate species. Professor Ruttner wants to clarify this contradiction. This requires crossbreeding tests. Western drones (*A. mellifera*) "should" mate with eastern queens (*A. cerana*) and vice versa. If there were reproductive offspring, eastern and western honey bees would belong to the same species.

Three years earlier, in 1966, there was an exchange of four bee colonies with a colleague from China. Four Chinese bee colonies, properly and very carefully packed, arrived undamaged at the airport in Frankfurt. Of these, two colonies survived. However, two bee colonies are not enough for successful crossbreeding experiments. There was

Fig. 18. Gudrun giving her talk at the Apimondia 1969 in Munich

no response from China to multiple inquiries for additional colonies. Obviously, the Cultural Revolution in China had also reached the universities in 1969 and international connections were no longer possible. At the University of Bonn, there is an entomological working group that is in active exchange with colleagues in Afghanistan. Professor Ruttner receives the promise that these colleagues would provide bee colonies of the eastern honey bee for Oberursel. That doesn't work either. The researchers return to Germany without bee colonies. They tell Professor Ruttner about an agricultural research Institute Tarnab in Pakistan, near Peshawar, a modern apiary of eastern honey bees. Professor Ruttner wrote to the head of the Tarnab Institute, a Dr. Rafik, and waited for an answer. Equally in vain.

However, Professor Ruttner had received research funding from the DFG for the crossbreeding experiments with *A. cerana*. His technical assistant Miss Agnes Bachler has been paid from this money for six months. Many other bills have also been paid with these funds.

"You can offer the DFG everything, including failures, but spending the money and not presenting any results–that's not possible! Niko, we absolutely have to get *A. cerana* quickly," says Professor Ruttner. "Can you help?"

For Niko, the request comes as a surprise. Of course, he wants to help. "What am I supposed to do?"

"Could you fly to Peshawar? In Tarnab, you have to visit Dr. Rafik and bring some bee colonies from there!"

Professor Ruttner has thought about this "assault" well. He knows that Asia, the "cradle" of honey bees, has played a central role in Niko's dreams for a long time. He thinks that Niko is easily seduced when it comes to exotic bees. Niko's answer is only partially positive:

"I'm no longer an independent student. I won't go without Gudrun's consent."

"You don't have to stay at home because of me and Anna," Gudrun answers Niko's question, "What are your chances of success? What if you don't get bee colonies? In the end, you have the money. Professor Ruttner will justify your travel expenses to the DFG. But a failure of the undertaking will ruin your reputation with the DFG. A high risk! But when I look at you, I'm sure you want to fly to Pakistan. And you have my blessing to do so."

Gudrun has no illusions. Niko's journey means an additional burden. Anna is only nine months old and constantly demands her mother's attention. And as soon as Niko has started, the proofs of the publications arrive, which she has to correct for many long hours in the evenings. Niko's publication in particular is a lot of trouble. Spelling is not exactly his forte.

Only three days after Professor Ruttner's question, Niko is sitting in a Lufthansa Boeing 707 on the flight to Karachi. A night flight, but hardly anyone sleeps. You quickly get into conversation with other passengers. "From where? Where?" and "What are you up to?" are the predominant themes. Niko gets important practical information about the country and its people. A German couple who has been working in Multan for several years as part of a development project of the Gesellschaft für Technische Zusammenarbeit give him useful information about the conditions in Peshawar and in the North-West Frontier Province. When asked which hotel in Peshawar is suitable for Europeans, Jan's Hotel is recommended. In Karachi, Niko changes planes and flies on with the PIA via Rawalpindi to Peshawar. The taxi at the airport in Peshawar charges 20 US dollars to the hotel. Much too expensive? Niko asks for a bus connection. Then a rickshaw, a scooter with three wheels, stops. He asks for the price. The young driver names a rupee amount. Niko has no rupees, only dollars, which the driver does not want or cannot accept. While he is thinking about how he can get rupees and the hotel, an older Pakistani dressed like an English gentleman approaches him: Can I help you? Niko thanks him for the friendly offer and asks for a bank to change dollars into rupees.

"Today the banks are closed."

Is that true? The Pakistani wants to help and would like to exchange 50 dollars for rupees for Niko:

"My exchange rate is more favorable than the official rate."

He will also take Niko to the hotel by car. Fifty dollars is a lot of money. Niko is dog-tired. He needs a bed and agrees. The car is an old British Morris Mini Minor. Niko's suitcase fits on the back seat and they drive off. He fell asleep during the journey? Nontheless, in front of the hotel, Niko gets a thick packet of rupees for his 50 dollars.

"No, I don't want to count. Thank you for your help," is all he says.

The first question in the hotel is about his reservation, but unfortunately Niko doesn't have one.

"All rooms are taken."

Niko pushes his passport across the table with some dollar bills and kindly asks if you can take another look? He will stay longer, and even a small, modest room is good for him. The receptionist takes the passport, asks him to wait and sends an assistant. "Fortunately," the review is successful. A valet takes his suitcase and takes it to a nice large room on the first floor with air conditioning and a private bathroom. Niko orders a large pot of tea, takes a long shower and then sleeps for many hours.

It's hot in Peshawar. In the early morning, the thermometer at the hotel entrance is at more than 112° Fahrenheit, which is 45° Celsius. As he leaves the comfortable hotel, Niko is overcome by the unaccustomed heat. He stops for a moment and breathes in the hot, humid air. Then the first slow steps. What happens there? Two men with rubber gloves and masks load a plastic bag with a motionless body into a city hearse. Another body lies in front of Niko on the side of the road. A passer-by, who must have noticed Niko's dismay, explains that many old people are fools and drink too little water:

"In this heat, they will die overnight!"

Niko is shocked. He turns around and hurries back to the hotel. In his pleasantly cool room, he throws himself on the bed. After about 10 minutes he has his composure again. He realizes that the difference between the two corpses on the street earlier and the numerous dead bodies that regularly appear on the public news on radio and television is primarily a question of personal distance. Will he be able to get used to the direct contact with dying and death? Niko is not sure!

The road to Tarnab leads past rice fields. On the right, Niko sees many beehives and a large tent. Are these the bees that the colleagues from Bonn have reported on? His taxi stops in front of the main building of the Institute. Niko pays off the previously agreed amount with the rupees exchanged yesterday. Some friendly young men stop and greet Niko and lead him to Dr. Rafik's office. There, in a large anteroom, a secretary sits behind a desk. A huge fan on the ceiling distributes the heat evenly in the room. Niko introduces himself and asks to be received by the chief entomologist. After a quarter of an hour, Dr. Rafik personally comes into the anteroom and invites Niko into his spacious office, in the middle of which is an impressive desk with three telephones. They take a seat in a seating area. Dr. Rafik asks about Niko's flight and in which hotel he is staying. They talk about airports in Pakistan and England. Dr. Rafik studied in Cardiff, "many years ago". Niko asks about the letters from Oberursel. Obviously, Dr. Rafik did not take the proposal and the offer of cooperation seriously.

"Such offers come here again and again. In most cases, however, the foreigners only want to sell their pesticides or equipment."

Niko assures him that Professor Ruttner is a respected university professor and does not want to sell anything.

"By the way, he would have loved to come himself. He has to give lectures at the University in Frankfurt, that's why he sent me. We have heard from German colleagues that you have made progress in honey production with native bees. May I take a look at the beekeeping at your place in Tarnab?"

Dr. Rafik replies: "No, the bees are just a hobby and not important. I have to take care of the rice cultivation. An introduced Philippine grasshopper has appeared here, which is very dangerous."

Niko was to come back the next morning. Mr. Shafi, Rafik's assistant, would then show him the bees. That ended the conversation. Dr. Rafik accompanied him back to the reception room and asks how Niko will get back to Peshawar.

"No, you won't find a taxi here, you will find buses." He talks briefly with his secretary. "And see you tomorrow!"

A car takes Niko to the nearby central bus stop. There are some beautifully painted minibuses in a row. A little boy sells him a ticket. The front seats are occupied. When Niko wants to sit down in one of the back seats, there are protests: "The back part is reserved for women. According to the laws of Islam, women and men must sit separately."

At the front, a few men move together and offer him the vacant seat. As soon as he sits down, questions come from all sides: Where does Niko come from? What could Niko want in Tarnab? Does Niko think Pakistan is good?

Foreigners seem to be rare here. Niko gets off at the central bus station in Peshawar and takes a rickshaw taxi back to the hotel. Over a large pot of tea, he thinks about how to proceed. First, he has to win the trust of Dr. Rafik. He will think that Niko has not just come to look at his bees. Unfortunately, Niko has no copies of Professor Ruttner's letters. Did he ask for bee colonies? Niko remembers the many fairy tales of Meir from Basra: "The older sons, who immediately and impatiently asks for the princess's hand in marriage, fail. Only the youngest son, who works patiently and waits until the old king asks him what he wants, wins the beautiful princess."

Niko doesn't need a princess, but needs bee colonies for Oberursel all the more urgently. He will put aside his "Western" impatience and wait until Dr. Rafik asks about his mission.

The next morning, Dr. Rafik introduces him to his assistant, Mr. Shafi, a young man Niko's age. He also invites him to dinner in his house, which is very close to the Institute's premises.

"Can you come around 6 p.m.?"

"Of course, thank you very much for the invitation. I'm looking forward to it!"

Mr. Shafi takes him to an Institute car in front of a garage. At the wheel of the jeep sits a tall driver in traditional clothing, a wide cape with harem pants and a loose headscarf. Niko asks Mr. Shafi why he doesn't drive himself.

"I am an Officer and Assistant Entomologist, as such I am entitled to a driver!"

Niko's answer that there are two cars in the Bee Institute in Germany, but no driver and that everyone has to drive themselves, provides a good introduction to detailed and long conversations about the people here and the people in Germany. Niko gets thirsty and asks for a cup of tea. They drive to the canteen and get a large jug, as well as two glasses of water. Niko doesn't dare to drink water. The tea has been good for him so far. Shafi is keenly interested in life in Germany and Niko's personal situation. He tells that Dr. Rafik and he belong to the Pashtun ethnic group, which is very important to him. He seems convinced that Pashtuns are the real masters on the ground. Dr. Rafik's family belongs to the most respected tribe in the region. His father had been the leader of a group of fighters and had fallen in the fight against the English occupiers. Rafik and his brothers had attended a college in Rawalpindi and then went to England to study. Obviously, the family is wealthy. An uncle, I suppose? Shafi does not know exactly. He himself has been married for four years. His wife came from a respected family and is distantly related to Dr. Rafik. He has two daughters and his wife has recently been expecting another child. Everyone hoped for a son. Shafi doesn't believe Niko that Anna, although only a girl, is very welcome: "Weren't you a little disappointed? Wouldn't you have been happier about a son?"

On the other hand, he can well understand that Gudrun and Niko have "found" each other independently of their families.

"That's probably one reason why so many marriages end in divorce in the West?"

"What about divorces here with you Pashtuns?"

Shafi claims: "We don't have divorces!"

Niko asks critically: "How is that possible? We humans are all the same!"

"The situation here is different. According to Islam, divorce is easy. I only have to tell my wife three times in the presence of witnesses: "I disown you!" and the divorce is done." Shafi laughs: "The groom has to bequeath all his possessions to the future wife as a sign of his true love in the presence of both families. In the event of a divorce, my shirt, shoes and all my possessions belong to my wife alone. She could chase me out of the house naked, no one would help me."

This ownership situation surprised Niko. He suspected a very extensive oppression of women. In Peshawar, all women in the streets and in public spaces wear a mostly black full veil, the burqa. In this heat, it must be torture! Niko prefers not to discuss the topic of religion and equality any further. Obviously, Shafi is a devout Muslim. Better to ask about the bees.

They set off in the direction of Peshawar. They stop at the apiary on the road, which Niko has seen on his way there. Shafi is greeted almost militarily. Two older men, presumably beekeepers, stand at attention in front of him. Two boys hurriedly come out of the big tent and stand behind the beekeepers. There are more than 20 beehives, Langstroth hives, which are set up in two rows parallel to a moat. Probably because of the high temperatures, large groups of bees are gathered at the hives entrance, which constantly ensure air circulation in the colony. In this way, the water introduced evaporates and provides cooling. Niko is amazed and enthusiastic. Because these bees fan exactly "the wrong way around"—the head points outwards, the tip of the abdomen points towards the entrance hole. So far, he has only seen fanning in the western *A. mellifera*: The head points to the entrance hole, the tip of the abdomen outwards! He cannot remember ever having read about this difference (Fig.19).

Niko asks the beekeepers about the foraging conditions. Shafi tries hard to translate his question into Pushtu, but obviously he doesn't work with the bees himself and doesn't know enough about beekeeping. This leads to sign language communication with the beekeepers, supported by Shafi, who then also participates. They laugh happily at misunderstandings. Niko's depiction of pollen collection and stinging behavior also triggers general hilarity. Step by step, Niko grasps the current situation: Because of the high heat, there are currently only a few flowers and very little nectar. The beekeepers feed the colonies with small portions of sugar syrup. However, robbery occurs again and again. They have lost colonies to absconding, which means that the bees and queen have moved out of their hive and flown away to look for areas with more nectar.

"May I please open a beehive?" asks Niko.

Shafi is unsure. Niko suspects that he would have needed Dr. Rafik's permission to do so. He withdraws his request and suggests going to Peshawar for lunch at Jan's Hotel. Shafi has to go back to the Institute, before he drops Niko off at the hotel.

In the evening, Niko is in Tarnab too early. At half past six he is standing in front of Dr. Rafik's large bungalow. You can't come too early! A short walk is a good way to pass the time. After a short time, Niko is called loudly. Niko doesn't understand what the two men holding guns in their hands want from him. In response to the questions that Niko does not understand, he repeats several times:

"Dr. Rafik Khan and Shafi Kahn!"

"You see Dr. Rafik?"

They hang their rifles over their shoulders again and accompany him back to the bungalow. From inside you can hear Dr. Rafik's loud voice. "Please wait". It seems very long to him. Then Dr. Rafik stands in the open door and greets him very friendly:

"Welcome to my house!"

He leads him into a large high room, in the middle of which is a large table, in one corner a stately television, armchairs, a sofa and small tables next to the armchairs. It is quite dark. The narrow windows are covered with dark curtains. A young man, whom Rafik introduces as his son, asks what Niko would like to drink.

"Would you like orange juice, Coca-Cola or water?"

Dr. Rafik explains that the provincial government has banned all alcohol.

"Has your suitcase been checked at the airport in Peshawar? Importing alcohol into the Northwest Frontier Province is not allowed! We follow the Koran! In Germany, people usually drink beer?"

"Yes, a glass of beer is customary with our meal. Personally, I don't like beer so much, but prefer to drink wine in the evening and when friends come."

Niko asks Rafik's son for Coke. Niko prefers to avoid water and juice—according to the general warnings in the travel guides. However, ice cubes float in the drink.

The shock of the armed guards is still in Niko's bones. He asks his host about it.

"Life here is complicated," Dr. Rafik answers. "There are enmities between the tribes and the families. You always have to expect that sooner or later someone will come and take revenge. That's why every settlement is well guarded."

"What are these acts of revenge about?" Niko wants to know.

"These are often hostilities that have existed for a very long time and that are continued again and again through murder or aggravated theft. Sometimes insults also lead to acts of revenge. We Pashtuns are a proud people. Sometimes perhaps too proud and not clever enough!"

Then Rafik changes the subject and they talk about the conditions at Western universities.

Rafik's son serves the food and asks them to come to the table. Rafik sits opposite Niko. There is chicken curry, rice, papadam and dal vegetables, plus tomato salad with green leaves he doesn't recognize. The curry tastes good, but is very spicy. Niko sticks more to the rice and the dal. After the ice cubes in the Coke, Niko now courageously drinks the water from the large glass in front of his plate. Rafik's son comes with a jug and refills it. After the meal, which Niko has duly praised and for which he thanks, they return to the armchairs. Rafik's son brings them biscuits and asks if Niko would like tea or coffee. Like Rafik, Niko asks for tea.

Rafik then asks how Niko found the bee colonies on the main road.

"The bees and the beekeepers impressed me very much. Everything looks extremely good. The place directly at the canal is well chosen in the current heat! The bees don't have to fly long to get enough water to cool their colonies!"

"Why are you interested in our bees? As far as I have read, bees in Europe and also in the USA produce much more honey than our bees. You have more than 100 kilograms per colony. Here in Pakistan, we are happy if we get ten to 20 kilograms!"

Niko is relieved and begins to explain: "We want to find out whether your bees and ours belong to the same species. If that's the case, you can create hybrids. You know that hybrids are often better than their parents? This is the case with plants and many animals. We want to research this with honey bees!"

Dr. Rafik understands him immediately: "You want to bring our bees to Germany for crossbreeding experiments? Transport by plane is certainly expensive! Do you have enough money?"

"No, we always have too little money, especially the universities in Germany have too little money. We received money through a grant for the crossbreeding experiments. Thus, air freight to Frankfurt is no problem. Before we talk about transport costs, I first have to get bee colonies! Could you help me?"

Dr. Rafik smiles and takes a sip of tea: "I have time the day after tomorrow and can show you the bees." Then he realizes that it is late: "My son will take you to the hotel in our car!"

Two days later, Niko meets Dr. Rafik at the apiary (Fig. 20). Dr. Rafik has a small decorated bamboo stick in his hand and talks to the elderly beekeeper.

Niko's joyful greeting is reciprocated reservedly, perhaps condescendingly. Does he want to demonstrate to the beekeepers that he is not impressed by a European?

Fig. 19. *Apis cerana* bees fan exactly "the other way around" compared to *A. mellifera*: the head points outwards, the tip of the abdomen points towards the entrance.

After a short conversation, Niko would like to inspect some bee colonies and asks:

"Which hives can I open?"

The beekeeper points to a box at the end of the back row (Fig. 21). The beekeeper asks Rafik if Niko needs a smoker and a bee veil.

"No, I'm careful and can work without a veil. A hive tool would be good!"

There is no hive tool, instead the beekeeper brings a short hook made of iron. Slowly, Niko lifts the hive lid. The bees react with a loud hiss. A surprise! The western bees do not hiss. Niko carefully loosens the combs with the iron hook. When Niko takes a medium comb out of the hive, only a few bees start flying. Niko finds older larvae and capped brood cells, but no queen. Niko hands the comb to Dr. Rafik, who stands behind him and watches closely. On the next comb, Niko finds eggs and then also the queen, which he wants to show Dr. Rafik. In doing so, he bumps into the comb that Rafik is holding in his hands. This is too much even for the gentle *A. cerana* bees and they start to sting. Niko quickly puts back the comb with the queen and takes the comb from Dr. Rafik, who stands behind him as if petrified. He has several bee stings and stinging bees on his face. The separation of the bee from the sting takes much longer with *A. cerana* than with western bees. They walk slowly under the canopy of the tent. Dr. Rafik sits down on a chair and Niko carefully removes the bee stings from his face. Rafik curses in English! Otherwise, he doesn't show anything. As the leader of the Pashtuns, you can't lose face even with many bee stings.

After this incident, things happen quickly. Dr. Rafik asks how many bee colonies Niko wants. He agrees with his proposal to initially transport four colonies. When Niko asks what the colonies cost, there is a usual answer: "Give me what you like." Then he asks:

"Could I get four good bee colonies from Oberursel in return?" Shortly afterwards, he adds: "If we want to do crossbreeding trials here, we also need an apparatus for instru-

Fig. 20. *Apis cerana* hives at an irrigation canal for permanent access to water. The tent in the background belongs to the beekeepers who watch the colonies day and night.

mental insemination of queen bees. Mr. Shafi could learn this technique from you in Germany?"

Niko agrees. This will work. "Please give me a letter to Professor Ruttner. I am convinced that the cooperation between Tarnab and Oberursel will be successful on both sides. Professor Ruttner is a reliable partner. You can count on him to send the bee colonies. And we are looking forward to the visit of Mr. Shafi. There will be no difficulty in getting a scholarship for Shafi."

Dr. Rafik is satisfied: "Niko, thanks for your support. I look forward to continuing our cooperation!"

Shafi and Niko have transport crates built in a carpenter's workshop in Tarnab. With a foxtail saw, hammer and nails, the boxes are ready the next day. The dimensions correspond exactly to the sketch! Everyone laughs and rejoices when Niko asks: "Where is the artist and engineer who built these perfect beehives?"

The return flight is booked. The four bee colonies are registered as excess baggage. PIA doesn't cause any problems. At Lufthansa, on the other hand, Niko cannot get a binding commitment for the transport of the bee colonies. Only the booking for his flight is confirmed by phone. During his flight to Karachi, Niko sits as if on hot coals. As soon as he has landed, he hurries to the baggage claim. Where are the bee colonies? He is referred to a quarantine station. All animals must be brought there. The building cannot be reached directly via the terminal. The address is written down for him. Niko takes a taxi there and luckily finds his bees before they have reached the quarantine room. Insecticides are regularly sprayed in such rooms to prevent the introduction of fleas, ticks and mosquitoes. This is a deadly environment for bees. They hand over the boxes to him and are happy to be rid of the loudly buzzing animals. The taxi driver refuses to take this dangerous load with him. Niko understands and pays a "hazard surcharge".

Fig. 21. Dr. Mohammed Rafik Khan (left), chief entomologist of Tarnab Research Institute and his assistant Mr. Mohammed Shafi Khan (right). Two beekeepers with bee veils wait in the background.

At the Lufthansa counter, Niko experiences German thoroughness. They look up guidelines, regulations and international regulations to see whether and if so, how bees are to be transported as additional luggage in the cabin. They don't find what they are looking for. The central office in Frankfurt cannot be reached by phone outside office hours. Then the boss has an idea how she can get rid of the responsibility for transporting the bees: "When transporting dangerous goods, the captain decides!"

An hour later, the pilot asks why the Bee Institute of the University of Frankfurt wants to get these bees. Niko shows the official import permit and remarks that these bees are much more peaceful than German bees. The pilot takes a close look at the transport crates:

"What happens if we encounter turbulence and a box breaks?"

"Then there are probably some bees flying around in the cabin. That's not much different than flies or bugs buzzing around. In addition, these boxes are custom-made. They won't break as long as the plane doesn't crash!"

"You are the expert I rely on you. We're taking these bees with us to Frankfurt!"

In Frankfurt, too, Niko only narrowly succeeds in preventing the bees from being taken to a quarantine room. Everything is prepared at the Institute. Professor Ruttner hardly has time to greet him. He is sitting with Mr. Blotz and the four transport boxes in the car on the way to Kronthal, where the *A. cerana* apiary has been set up.

Gudrun picks up Niko with Anna. Niko gets in the back and sits down next to his daughter Anna. She beams at him and wants to sit on his lap. She tries to talk. She learned her first words in the 18 days he was on the road. Then she falls asleep. Niko carries the sleeping child into the apartment and lays it on the bed. Gudrun reports on the "terrible" time without him. Niko agrees with her:

"Neither of us are good lone fighters anymore."

He can't answer Gudrun's questions: "Now I have to sleep first. Tomorrow I'll tell you everything in detail."

Twenty-fourth chapter

Gudrun and Niko have their second child, move and are happy to have visitors

During the next bee season, Gudrun's life changes from the ground up. At the end of May, their son is born. He is given the name of their friend from Israel, who of all things bears the less Jewish name Fritz. Now Gudrun can no longer think of research. Fortunately, it is much warmer at this time of year than when Anna was born at the end of December, only in the kitchen Gudrun has to heat up the coal stove for bathing, which provides hot water and a warm room at the same time.

A short time later, Professor Ruttner manages to get a regular assistant position for Niko at the University of Frankfurt for the Bee Institute. This is a great relief for the young family. The assistant's salary is almost twice as high as the DFG scholarship. In addition, Niko is now entitled to an official apartment. The university has the right of occupancy for several apartment blocks in the Limes city near Oberursel. In September, an apartment on the 3rd floor will become available. Then, after eight years, they will have their first apartment with a bathroom. There is running hot water in the kitchen, and instead of coal stoves there is central heating. The move this time is a bit more difficult because of the furniture that has been purchased in the meantime, but with the Institute bus, friends and students it is done in one day.

The Institute is lively. To Niko's great joy, Mrs. Alpern, their hostess in Israel, arrives. She received a scholarship from the DAAD (German Academic Exchange Service) and is now learning instrumental insemination of queens in Oberursel. Full of joy, Gudrun and Niko invite her to their home and talk about their time together in Israel. They learn that at that time the invitation of Germans had led to a dispute in the ministry. Mrs. Alpern and many others also had reservations about bringing in German students. Today, everyone is convinced that it was the right decision. At the moment, two German students from Frankfurt are working in Zrifin again.

Turkish colleague, Alef Settar from the University of Izmir, has come with a large pile of bee samples from different regions of Turkey. A morphometric analysis is to be used to determine the different breeds and ecotypes. This geographical area in Turkey is very important to understand the diversity of European honey bees. Professor Ruttner is thrilled that these new bee samples can finally close a gap in his bee collection. Niko's admiration is above all for Alef Settar. For each bee, 32 characteristics must be measured and each sample comes from a colony and includes at least 30 bees. He has brought a little more than 80 samples, i.e. more than 2,400 bees. If he measures 32 traits for each bee, that's more than 76,800 individual traits. Niko gets dizzy when he thinks about this workload. From early in the morning until late at night, Alef sits behind his microscope and works with perseverance and determination. Only rarely does Niko succeed in persuading him to take a break or visit Gudrun.

With Professor Jakov Ishay from Tel Aviv, wasps and hornets move into the Bee Institute. He investigates the important role of wasp larvae in the social life of the colonies. In emergency situations, when the wasps cannot find food, the weight of the larvae decreases. Niko and Gudrun watch with great interest and somewhat puzzled as the hungry workers milk their larvae. Apparently–according to Professor Ishay–the larvae are not only living reservoirs, but can also biochemically transform or convert the animal food fed to them. In habitats such as the deserts of Israel, the hornets mainly hunt other insects or eat carrion. Their prey consists mainly of protein, which is fed to the larvae. However, the secretions that the larvae release to the adults contain carbohydrates, especially dextrose. According to Professor Ishay, the formation of D-glucose from organic non-carbohydrate precursors is the key that opens up desert habitats for wasps where there are no plant juices or other sources of carbohydrates.

There are no deserts in Oberursel. Thus, a large flight cage is built for Professor Ishay, which is colonized with a large hornet colony. The animals are provided with chemically defined feed according to experimental projects. Professor Ishay then "milks" the larvae. To this end, he has designed and constructed a special cage enabling him to lock out the adult flying hornets. This obviously does not always work, because one day loud screams can be heard in the Institute. Prof. Ishay is lying on the ground in the hornet cage! Niko runs to get his bee veil, then protected with a veil he enters the cage and helps Ishay to his feet. His right hand is injured and bleeding profusely. Niko supports him and slowly, surrounded by curious but peaceful hornets, they go outside. Professor Ishay has not been stung; the deep cut in his hand stems from a broken test tube that he was carrying when he fell. He is immediately taken to the emergency doctor and so he does not hear the comments of the beekeepers and students. For Professor Ruttner, this is an opportunity to demonstrate that bee or hornet stings are less dangerous than a panicked escape:

"By the way, bees and wasps fly faster than even the atheletes

can run: So just stop! Walking away very slowly is the better choice."

Overall, Niko is grateful to Professor Ishay. He learned a lot of new things from him. Professor Ruttner is also satisfied: wasps will complement research at the Institute in the future. As an alternative to honey bees, the individual animals in these species are much more independent. The wasp colony is easier to understand than the bee colony.

Niko is fully occupied with teaching and research. There are still many demonstrations at the University, especially in the house-to-house struggle for the beautiful old villas near the Zoological Institute in Frankfurt's Westend. They are to be demolished and replaced by modern concrete office buildings. The methods used to evict tenants are drastic. Necessary repairs are deliberately not done, and finally the houses are evacuated for ostensibley safety reasons. At the same time, there is still a great shortage of housing. This creates a protest movement that Gudrun finds important and right, but in which she cannot participate with the two children. For example, a few hundred meters from the Zoological Institute, a beautiful old villa that she knows is empty. Many students move in there.

Gudrun learns more about this from the radio than from Niko although he passes the house on his way home. When he comes home in the evening, he enjoys playing with the children and putting them to bed. After that, there is little time to talk about research and protests at the University. Gudrun is also usually dog-tired. And her experiences with the other mothers at the sandbox or the quarrels between the children are not worth mentioning.

Sometimes Niko brings colleagues home with him. For Gudrun, this is a welcome opportunity to keep in touch with the world of science. Gradually even that is becoming increasingly difficult. The guests are happy about the lively children and prefer to play with them rather than discuss experiments. They advise Gudrun to read scientific literature regularly, but this is not easy for her. The children interrupt her again and again. Small misfortunes lead to tears, they need Gudrun's comfort. Or something to drink, something to eat or want to play with her. And why should Gudrun continue to deal with publications at all? She doesn't want to carry out any more experiments, does she? Gudrun had resolutely left bee research behind because of the children. Nevertheless, more and more often the loss of her own scientific work hurts her. While she looks after Anna and Fritz at the sandbox, Gudrun tries to talk to other mothers about giving up her own job. For many women, this is not an issue. They speak of "mother's happiness". And above all, there is a discussion about what they cook to "pamper" their husbands.

Gudrun had longed for the children and had been looking forward to family life with Niko. The thought that she would miss bee research hadn't crossed her mind at all. Aren't your own children more important than bees? What's wrong with her? Has the scientific work spoiled Gudrun for a lifetime limited to mother and housewife?

Twenty-fifth chapter

Niko again fetches bees from Pakistan

Then Niko comes up with the question of whether Gudrun agrees that he should fly back to Pakistan. This time for six weeks. There are difficulties with the Pakistani bee colonies. The colonies from the hot, subtropical Tarnab fare much worse with the conditions in Oberursel than the formerly imported colonies from China. This time, Niko is supposed to try to get bee colonies from the mountains in the border area with Afghanistan. Professor Ruttner hopes that the bees from the colder mountain regions of the Himalayas would survive the winter in Germany better.

"This clever Ruttner wants to send you again," Gudrun says indignantly: "For him a convenient and cheap solution! Without his own risk! Have you thought about who will benefit from your trip? Professor Ruttner determines, and perhaps also Professor Woyke, whom he wants to bring on board for the crossing experiments. And what is left for you?"

"You're right, Gudrun. Professor Ruttner thinks of himself first. For me, the main goal of this trip is to gain data and information for my own research proposal. In addition to *A. cerana*, I will also be able to observe the giant and dwarf honey bees. Then I can drive to Pakistan with you and the children next year and study Asian honey bees in their natural habitat in Tarnab for a few months."

"Niko, keep your feet on the ground! With two small children in a jeep through Iran, through Afghanistan to Peshawar? You're crazy!" Gudrun is annoyed: "At the moment it's all about the now. I have to stay here, alone with both children, for more than six weeks. I'm on my feet from six in the morning to eight in the evening. And if you don't take the children in the evening, I won't have any peace at all. That's the situation we're talking about."

Gudrun understands that Niko wants to go back to Pakistan. Before they had children, they always dreamed of adventurous journeys. And during the discussion about whether they should start a family, Niko had argued that he could well imagine research, travel and adventures with Gudrun without children. Gudrun was the one who preferred to stay at home with children. At the end:

"I agree. I'm also afraid that it can be dangerous for you."

"Gudrun, believe me, this time everything is much easier than on my first trip to Pakistan, which was like jumping in at the deep end. Dr. Rafik in Tarnab himself suggests that I should come for six weeks and offers me a small bungalow next to his Institute as accommodation. He also wants to support my excursion into the mountains. For gasoline costs, he can give me a Landcruiser. In addition, Shafi and a second officer are supposed to accompany me the whole time. Dr. Rafik has probably understood that we are serious about cooperation after he got bee colonies and the apparatus for instrumental insemination."

Niko starts his second trip to Pakistan (April 19, 1973) in a good mood and full of confidence. He has his eyes on his future. The Asian honey bees offer exciting research opportunities off the beaten track of western *Apis mellifera* research for many years. He is sure that it will not be his last trip to Asia. Perhaps the last journey on behalf of Professor Ruttner. In the future, he wants to travel with his own research project and his own funds.

Almost every day Gudrun receives a letter from Niko, sometimes two or three at a time because of the irregularity of the mail. And she holds down the fort for six weeks. Unfortunately, the children soon get sick and she has to take care of them at night. Every minute that both children sleep at the same time, she sleeps too. She cannot maintain contact with her friends who do not yet have children and are working.

She is very happy when Niko finally comes home. Gudrun drives to the airport and they are in each other's arms again. After the first greeting and the loud joy of the children, Gudrun says: "Never leave me and the children alone for so long again."

"Yes, the six weeks without you and the children were far too much. I will take the three of you with me in the future."

Gudrun looks at him doubtfully. Niko is emaciated and looks sick. They drive home and after a small meal he goes to sleep, Niko sleeps for 14 hours.

The next morning, Anna proudly shows her father her new treasures and towers, which she built herself from Legos. Niko praises her building skills and wonders what she has learned in six weeks. Fritz, now almost a year old, also crawls to his teddies after enough cuddling and wants to show Niko everything. Gudrun has to wait.

Professor Ruttner calls and Niko has to go to the Institute. He is back after only three hours and looks very miserable.

"Niko, what's wrong with you?" Gudrun asks him in the evening. "You have to go to the doctor. Have you been sick? You didn't write me anything about that. You always say you're fine, even if you're terminally ill!"

"Oh, Gudrun, one after the other. In the beginning, everything was as easy as hoped. Mr. Shafi was waiting for me at the airport in Peshawar. A warm welcome! No taxi and hotel problems like the first time. He proudly accepted the plush elephant for his little son and my congratulations. The meeting with Dr. Rafik was also good, he immediately introduced me to his new colleague Mr. Abdul Quayum, a young man about my age. He comes from the Punjab, as Dr. Rafik remarked rather condescendingly. Quayum is much smaller, stockier and darker-skinned, so no Pashtun. He helped me a lot. My bungalow turned out to be a room in an elongated wooden barrack with many residential units. There was an old bed with a dirty mosquito net with holes, a table with two chairs, a cupboard and an adjoining room with shower and toilet, all much dirtier than back then in Israel. Quayum helped me buy everything I needed in Peshawar. Thanks to his negotiating skills, I only paid 50 percent of what I would have paid without him. He has also arranged a daily "local breakfast" for me, which is flatbread with lentils and egg."

"It also reminded me a bit of Israel. You needed Quayum's help even more urgently than we did back then. After all, we had our motorcycle and despite our isolated location, we were mobile and, above all, independent."

"Yes, it gets even better. The next morning, Quayum showed me a eucalyptus tree in which a large colony of the giant honey bee was hanging (Fig. 22, Fig. 23). As you know, I've been wishing for a long time to finally see these bees in real life!"

"Hopefully you've protected yourself well! I remember the firefighter equipment that Professor Morse was talking about." Gudrun has to laugh at this memory.

"Of course, I put on the full American beekeeper's protective gear, in spite of the heat, even gloves. Otherwise, I wouldn't be sitting here today. We very carefully leaned a ladder against the trunk of the eucalyptus tree and attached it with ropes. Then I climbed up to one and a half meters from the colony–in slow motion, of course! The bees didn't let themselves be disturbed."

Niko is still beaming at the memory:

"You should have seen that too, I'm sure you'd be just as excited as I was. There was nothing to be seen of the comb itself, it is enclosed by a curtain of workers who hang down from the branch in strict order with their heads up. All flying bees land and take off at a clearly defined place at the bottom of the comb, at a point where a ray of sunlight hits the comb. There, the regular pattern of the bee curtain is interrupted, a take-off, landing and dancing place, so to speak. I felt quite safe when suddenly there was movement at one point on the bee curtain. The bees swung their abdomen quickly back and forth. At first, I thought they had discovered me. Then it occurred to me that it is shimmering that has been described by Butler and other authors. Then I saw a small fly flying up and down close to the curtain of bees. The bees closest to this fly reacted with shimmering. Later I saw a butterfly that was also flying back and forth in front of the nest and its approach triggered shimmering locally at the prospective landing point. This will probably prevent the landing!"

"This must be really great to watch! When the children are grown, I want to see that too."

"I'm sure I'll show you! You always have to be careful. I then made a very stupid mistake. I asked Quayum to get tea for us. As soon as he left, a group of little boys arrives. They had probably been watching us for a long time and now, without Quayum they dared to come closer. They were curious: a strangely hooded European on a ladder with the bees—you don't see that every day. The boys stopped and called out their friendly 'hello' to me. I tried to send them away with 'Go Away'. That only triggered laughter. Even more children came. Then I became scared—how long would the giant honey bees tolerate the disturbance of the children? That's probably why I descended the ladder too quickly. I was promptly stung. The children knew these bees and their stingers. As soon as they saw the first stinging bees, they were gone. I, too, had no choice but make a quick and orderly retreat. The stinging defenders accompanied me for a surprisingly long time—more than a kilometer. I realized: These bees must have better lasting alarm pheromones than the western *A. mellifera*. I want to, yes: I have to investigate this later."

"You are incorrigible, instead of taking it easy and thinking about how to get to the mountains, you think about new experiments! No wonder you look so thin and pale!"

"Gudrun, the whole story ends sadder than you think. When Mr. Quayum came back at last, and saw how closely the bees circled and stung me, he called to me from a distance, "We will meet you in front of your room." As soon as I was in the house, the bees did not follow me. I was quite shocked by the incident. How easily could the children have been harmed? Mr. Quayum reassured me: "Here, almost every boy has thrown a stone at a *dorsata* nest and then been stung," he said. "Every child here knows that."

Niko pauses: "Unfortunately, he wasn't right". After some children were stung, Dr. Rafik had the *dorsata* colony killed. He was even proud of the fact that giant honey bees didn't stand a chance against his power spray. In addition, the people were satisfied. The bees were dead and they got almost 10 liters of honey. Imagine: They ate it, even though a lot of poisonous insecticide had been sprayed on the honey. That was all my mistake! I didn't know that as a white-skinned European, you attract the children in this remote area like a magnet!"

"That's really sad. You should have waited and prepared everything carefully!"

"Oh Gudrun, it was clear from the beginning that Dr. Rafik would not organize my trip to the mountains until the following week, i.e. only when I was really with him. Everything is slow there. You have no idea how it really went. I'd rather not write you that exactly. We were to go by jeep to the Swat Valley to Mingora, together with Shafi and Quayum! Shafi knows some beekeepers there, who would then take me further into the mountains in the direction of Gilgit or Chitral—into the free tribal territory of the Pashtuns. Mr. Shafi, who is familiar with this area, had not met any beekeepers in the high mountains, but he had seen some colonies in the settlements, most of which nest in the walls of the houses. That's where we wanted to go. Gudrun, I'm dead tired again. Let's go to sleep, tomorrow evening I will tell you the rest!!"

The next day, Niko calls the Institute of Tropical Medicine and asks what samples they need to test him for diseases from Pakistan. He describes that he probably had amoebic dysentery and perhaps other infections. He makes an appointment with his family doctor and drives back to the Institute. Gudrun is relieved that Niko wants to be thoroughly examined. As before the trip, it is almost evening when he comes back home. Anna and Fritz are happy and have a lot to tell and show him. After he played with them and cuddled them extensively, they fall asleep contentedly. Often Niko falls asleep with them. Gudrun has to wake him up. Finally, she has Niko for herself:

"I want to know the truth about your journey to the tribal areas."

Niko wants to make himself comfortable first, Gudrun gets two glasses of wine.

"You'll hardly believe what I'm saying now. I'm not exaggerating. Even the preparations for our trip to the Swat Valley surprised me. Shafi asked me if I could handle weapons. No problem, I answered, I practiced as a soldier with rifle, pistol and submachine gun. I'm a pretty good shooter. Why do you ask?

Mr. Shafi handed me a pistol and pointed to a submachine gun in the cupboard: "We will need these weapons," he said. In the tribal areas there are no police, no military, only free Pashtuns, and only a part of the population adheres to the general right of hospitality. The people are very poor. There are always raids and fights between individual groups. That is why we must also be prepared for highway robbers".

No doubt, A foreigner with a gun in his hands would certainly be the first target of attackers.

"I won't touch a pistol or any other weapon," I said. "I am a peaceful traveler and pose no danger, that is the message I want to convey. I will come unarmed and rely on you. We must not take any risks that would jeopardize the success of our mission. The procurement of bee colonies is important. Risk our lives to do so? No, I don't want that."

Shafi sensed my fear and tried to calm me down: "We have helped many families up there to protect their crops from pests," he said. "I have been there many times and you will see; we will be welcome there and treated like important guests. We will be protected and everyone will help us. Until we get there, we have to show publicly that we can defend ourselves." I realized that Mr. Shafi is a proud Pashtun. He knows from experience how to behave in the tribal areas."

Gudrun shakes her head in disbelief: "Where on earth could we have read this before? Not in normal books. These conditions are described as if they were in times long past." Niko nods: "In any case, I was not aware of this dangerous situation when I took on this assignment in Oberursel. I thought that everything was like in Tarnab, just a little more primitive. Sleeping in a tent, not enough to eat or the like. I didn't prepare myself thoroughly enough. I was interested in the transport possibilities of bee colonies, not in the living conditions in the border area between Pakistan on the one hand and China and Afghanistan on the other. That was a mistake that won't happen to me again." Gudrun toasts Niko with her glass of red wine:

"You won't believe how glad I am that you're sitting across from me, even if emaciated." Niko also raises his glass: "You're not the only one who is happy. It continued with many

Fig. 22. A ladder fixed to a tree near an *Apis dorsata* colony.

Fig. 23. A dense curtain of worker bees covers the single comb of *Apis dorsata*.

surprises. In addition to the weapons, two automatic rifles, two pistols and a submachine gun, we packed four boxes each with Coke and water, my cream crackers and all the bee equipment into the jeep. There was also a large package of provisions that Shafi's wife had packed and his personal luggage, a small suitcase, and my travel bag. Dr. Rafik said goodbye to us in the evening because we left very early the next morning. We had two drivers with us, both also Pashtuns. When I asked about Mr. Quayum, I was told that he was a Punjabi and was just a burden!

At first, things progressed smoothly. Then there was only a narrow asphalt road with sandy, extended shoulders on the left and right. Our driver seemed to be completely out of his mind. Instead of slowing down, he raced at high speed towards an oncoming car. The other car didn't seem to brake either. I dove behind the driver's seat only a fraction of a second before the collision. Then I heard roars of laughter and shouts of joy. Our jeep continued on the asphalt strip in the middle. The oncoming car had swerved into the shoulder and seemed to have skidded violently. Because of the clouds of dust, I couldn't see if it had gone well. I protested loudly and wanted to stop such Russian roulette. Shafi talked to the driver, he answered angrily.

Shafi explained to me: "These drivers know exactly what they can and can't do. They want to ride the way it's right. They don't want advice from foreigners."

I couldn't shake the suspicion that Shafi was happy about this answer. Pride and the feeling of superiority seem to be a basic attitude of the Pashtuns. All I could do was assure him that I appreciate his driving skills and trust that he would bring us safely to our destination. "Maybe we can slow down a little so I can see the great scenery and mountains better," I added in a friendly manner. This apology was well received. In any case, they drove slower for the next few kilometers."

"My goodness, I would have gotten out and not driven a meter further with these people. I have no understanding for something like that!"

"After a few hours, we had to stop at a post. Shafi talked to the soldiers and then explained to me: "This is the end of Pakistani-controlled territory. We are going to an autonomous tribal area." Shortly afterwards we took a break in a settlement. Two elderly men came to our car and greeted us. Shafi gave everyone a bottle of Coke. We looked around. There was a shop and stalls with vegetables, melons, onions and more. On some flowers I observed *A. cerana* collecting pollen. Feral, free-living bee colonies? At least the people there knew nothing about a beekeeper or beekeeping. Shafi showed me a workshop where weapons were made. I took a closer look at a submachine gun. An Israeli Uzi? I knew the model from my military service. Not an original I asked myself. Shafi explained: "The people here are highly skilled craftsmen and earn a lot of money. The weapons are smuggled all over the world and fetch good prices." He looked at me questioningly. For a bargain price of 100 US dollars, I didn't want to buy this gun.

Fortunately, there was not only the gun shop, but also small eating establishments. The food was served outside under a canopy. The kitchen consisted of a barbecue and a wood fire with sooty pots. Anything but hygienic. The food tasted good. And the tea was better than in Tarnab. I praised the food. Everything was smuggled, Shafi said: no taxes, no customs duties. A free life. I pointed to the numerous children in rags and in obviously poor health. Isn't that a disadvantage of this free life? He reacted indignantly. "When we are back," he said, "I will show you the misery in Peshawar. Compared to the slums in Pakistani cities, this is a paradise for children. Here people live according to the Koran and everyone gives from their possessions to the poor. The school, a large, half-open building, is also paid for by the voluntary donations of the rich, including the mullah who teaches here." I didn't want to continue the discussion. The certainty that Islam has the right answers to all problems is traditionally firmly anchored and has shaped life and survival in these harsh mountain regions for centuries."

"Have you caught amoebic dysentery there yet?"

"No, it was all cooked or well fried! For the onward journey, a young man accompanied us with a rifle slung over his neck and two pistols on his belt. He was introduced to me as our guide, but he was probably first and foremost our watchdog. I had the impression that the visit of a German aroused suspicion. Shafi and I believed that the purpose of our journey was only bees. From then on, we drove off the main road on unpaved roads. Our destination was a settlement that is famous for beekeeping and honey. Steep mountains and very narrow valleys with small plots and fields where poppies bloom. I asked for a short break because I wanted to see if honey bees were visiting the poppies. After a short discussion with our guide, Shafi refused my request.

"If we stop here and you go to the poppy field," he said, "it could lead to misunderstandings. We are probably being watched by the people here. If you want to see a poppy field, we should ask in the next settlement and then inspect a field with its farmer." This argument convinced me immediately. Shafi explained to me that up to the mountains of Afghanistan, poppies are grown to produce large quantities of opium. "Our government has distributed seeds from Mexico of new wheat varieties to farmers free of charge," he continued. "This has been successful. Unfortunately, these plants are susceptible to caterpillars and other pests. I helped the farmers in the Swat Valley with pest control. That's why I know my way around here so well." Why are there still such large opium fields here, I asked. "Of course, the cultivation of poppies provides the economic base but the new wheat improves the local food supply; flour and bread have become cheaper and are of better quality."

The settlement, which we reached in the late afternoon, consisted of less than 20 buildings, which were built into the slope to the left and right of the small river. Shafi was warmly greeted by an older man with a long beard. When asked about bees, the old man said: Yes, there are some bee colonies. They would like to show them to us, the bees are further up in the gorge. "You can't get there by car," he explained. With a guide, we would be able to reach the bee colonies on foot, it would take more than two hours. Now it is too late for that.

Shafi told me that we should definitely stay here and look at the bees tomorrow. Of course, I agreed. "Before we go to the bees tomorrow, we have to buy a bee colony," I said. "I want to take the queen with me, and for that I have to look for her." Shafi wasn't sure if we could buy a bee colony. At dinner, the owner of the bee colonies came. He was unsure because bee colonies were otherwise not traded there.

Shafi explained the situation to me: "Bee colonies belong to Allah, who sent them. The swarm of bees is a sign of Allah's blessing. The honey itself belongs to the beekeeper or the owner of the house. Usually, the honey is given to a Hakim. Or to a pregnant woman or to a sick person". I explained to him that I wanted to give the honey from the colony to the beekeeper. I would leave everything there, only the queen with ten bees would be needed. If everything went well, the bees would raise a new queen and the colony would survive. "I definitely don't want to take the queen with me without the owner's consent," I stressed. "We have to have permission tomorrow morning. If not, then we'll move on." Shafi agreed, only the price question seemed to be a problem.

"Mr. Shafi," I said, "that's for you to decide. The seller must not feel that he has been treated unfairly. Especially here, where everyone has a weapon!" Shafi thought about it and suggested that we could use the price of a good chicken or a fat rooster as a guideline. However, the price of a goat, a sheep or a mule as a comparison is exaggerated.

I was tired and decided to leave the negotiation to Shafi. I had to go to the river to brush my teeth. In the guest house there was a large bedroom with two-storey wooden beds with dirty old mattresses and woolen blankets. There was no other accommodation. Like our drivers, I went to sleep dressed. The feet were secured with socks, the face with a towel against mosquito bites. In the morning, contrary to expectations, there was a good breakfast: hot tea and toasted bread, plus two fried eggs. Shafi then came. "Good morning, good news. Last night I agreed with the beekeeper on the price for the queen bee. We pay five US dollars, twice the price of a good rooster." Everyone agreed with the chicken price, but they said that as a German I had to pay twice the price. I had expected much more. I was happy because we now had a good starting point for further price negotiations."

"That doesn't sound so problematic," Gudrun said. "Negotiated a price for bees. The mosquitoes without a mosquito net were certainly unpleasant, you still had anti-malaria tablets. When did you get so sick?"

"I'll tell you that, but then I have to sleep." Niko yawns. "Well, we finally set off into the mountains. Our luggage, I packed a smoker, smoke material, bee veil, cages for queens and other beekeeping utensils, everything was loaded onto a mule and then we started walking. Steep uphill, and the pace was quite high. After an hour, I had to ask for a break. I was thirsty and needed a sip of water. We, however, hadn't come with our water bottles! There was only one water canister from the village. What was left for me? I drank from the cup offered."

"According to travel guides, this is a mortal sin," Gudrun interrupts him. "I had to take the risk. Drinking too little is deadly at these temperatures. I experienced that on my first trip. After three hours we finally reached our destination. An abandoned old house on the southern slope of a field of loose stones. The bees nested in a thick, jagged stone wall. There were probably at least six colonies. I found six flight holes with busy air traffic. The beekeeper said that there were many more colonies.

"Are the bee colonies here always in houses, walls or other structures built by humans?" I asked. Shafi translated the beekeeper's answer.

He said yes. Most of the time, when they build a wall, they would leave a niche for the bees. There are no hollow trees up there like down in the Indus Valley. Bees and humans should live together in the mountains!

That was new to me. Obviously, the availability of nesting cavities is an essential condition for the proliferation of *A. cerana*. In the course of this excursion, I decided to ask further questions and find out whether there were also natural nesting sites in the high mountains that were independent of humans. Now it was more important to look at the colonies. I asked the beekeeper how he took the honey from the colonies. He wanted to show me. I showed him the smoker and lit it. Then I spread out a cloth and put on my veil. The beekeeper removed a stone slab in front of the colony. Many bees flew off and circled us. The people quickly left, only Shafi stayed with us. I gave him the smoker, which he used carefully and prudently. The beekeeper carefully loosened the outermost comb with a long knife and slowly took it out without crushing the bees still sitting on it. I carefully placed the comb on a clean cloth. The following comb was also full of honey. On the third comb there were brood cells. The beekeeper gave me the knife and carefully I removed the following combs from the walls of the nesting cave. The young bees stayed on the comb and gathered in a dense cluster. No queen was seen but on the third comb did I found the queen and carefully put her in a small cage. Then I added six workers and as provisions a small portion of dough made of honey with powdered sugar, which I always had with me as a precaution. I placed the cage into my breast pocket, it was dark and warm there. Carefully I mounted the brood combs back into the nest cave. The beekeeper closed the colony with the stone slab. He packed the combs in a plastic jug. Then the beekeeper sat down in front of the colony and prayed. Shafi also took part in the prayer. After that, we immediately made our way back. In total, the search for the queen had lasted almost two hours.

We were in a hurry to reach the settlement before nightfall. I was totally exhausted and after a cup of tea I went straight to sleep. During the night, I was woken up by a stomach ache. I swallowed a German Mexaform-S for diarrhea. That didn't help. In the morning I still felt bad, the stomach pain became more severe. Shafi looked at me worriedly and asked if we could leave. "Yes," I answered, "I need a doctor or a hospital." Shafi decided to go to Mingora. On the way, my condition deteriorated. During a break, I noticed blood in my stool.

Shafi made a diagnosis: "This is certainly amoebic dysentery". Not very reassuring. I don't remember how we arrived in Mingora. It was only in a small traditional hospital that I regained consciousness. I got a lot of pills, probably from China, and they helped. The fever was gone, but I was slow to recover. Two days of bed rest were necessary before we could set off again. I was still too weak to endure long walks. While I slept in the car and drank more than two liters of tea a day, Shafi took over the marches to the beekeepers. In total, he then brought two more queens, so that after twelve days we had reached our destination and could return to Tarnab with three high mountain queens. I put them in three small colonies without brood. In two days I saw many bees with pollen. This was a sign that the queens were accepted by the bees, newly laid eggs so now the bees collected pollen to raise the larvae."

Niko yawned and Gudrun too. It's almost midnight and at five o'clock in the morning the night's rest is over.

At breakfast, Gudrun can't wait any longer. "Didn't you say at the airport that you wanted to take the children and me with you on your next trip to Pakistan. Do you really think that I want to work with you at 40 degrees in a long black dress with my face covered with a thick veil? Or want to be locked up in a courtyard with the children all day? And I don't want to see a severed hand in the market and hear the remark: This hand will never steal a sheep again! You had written that to me in the letter! Do you want our children to be exposed to the diseases there? No. I will not go to such a country! Neither with or without children! You don't even have to ask."

Anna and Fritz were scared. They are not used to Gudrun speaking so loudly. Anna starts crying. Gudrun is frightened and Niko tries to calm the children down. Before he disappears into the children's room with them, he says calmly:

"We'll talk about that later! Of course, I don't want to go to the tribal areas with you. I'll tell you about the experiments in Tarnab later!"

There is no other way, Gudrun has to wait until the children are asleep to talk to Niko again. Finally, they sit comfortably in the living room with a glass of red wine.

"Gudrun, please let me finish my experiences first. Only when you know everything does it make sense to discuss traveling together. I don't want to get any more bees; I want to get to know the dwarf and giant honey bees in their habitat. And that works very well in Tarnab at the Institute. Because as soon as I was back there, my health was much better. I never got sick there."

"Okay. Your description of the first *A. dorsata* colony was not easy either."

"You're right. Believe it or not, I'm capable of learning! I was more careful when observing the new colonies. I observed them from a distance and even with binoculars. In the beginning, I ran away unnecessarily twice. The first time, the order in the bee curtain suddenly dissolved and it looked as if it was seething. Then suddenly thousands of bees started in my direction, as I thought and I withdrew. Later I found many yellow drops of bee feces on leaves and also on the ground: it had only been a cleansing flight." "It was Yellow Rain, which the U.S. Army took for poison rain in the Vietnam War. A bee scientist, Tom Seeley, explained to them that it was only bee droppings."

Niko laughs: "That's really a lot of droppings! It took me two or three days to better understand the communication of these bees. After that, I didn't need a veil or other protective clothing anymore. An orderly bee curtain shows that everything is safe and that the colony does not perceive any threat. Incipient disorder in the curtain combined with sibilants, a hissing sound, is a forewarning. If the shape of the colony changes, i.e., the curtain becomes thinner and many bees form chains under the comb that look like long beards, then you better disappear. These alarm signals initiate the dreaded social defense behavior. During the defense, the beard of bees suddenly falls off and suddenly thousands of bees are in the air, which helps to drive away the enemy. In any case, there is still enough time from the beginning of the chain formation for a safe retreat. These bees are actually busy all day long and—as a night observation revealed—even at night in the moonlight, warding off honey robbers and enemies. A considerable effort, especially compared to the cavity species, who are protected all around by the walls of the nesting box and only have to guard the entrance hole."

Gudrun interrupts again: "It's really a huge effort that the giant honey bees have to make. A living all-round protection! Why don't they breed in caves like our honey bees?"

"I asked myself that too. What advantages do free-living bees have that the higher defense costs are worthwhile? My explanation so far is that caves are needed by a lot of animals of many different species. By dispensing with nesting holes, *A. dorsata* has an almost inexhaustible selection of nesting sites."

Gudrun rocks her head back and forth: "That sounds plausible and the giant honey bees are large and have long stingers. What about the free-living dwarfs? The colonies consist of far fewer bees and their stingers are very short! Do they have a worse poison?"

"No, not that I know of! The dwarf honey bees hide their nests in the dense branches! They are not so easy to find."

"You also wrote me something about the dwarf honey bees."

"Yes, Quayum showed me an *A. florea* nest. It was built around a thin branch and looked like a huge leaf from a distance. A gardener had found it while cutting back a bush."

"You mean the giants threaten with their size and the dwarfs hide?"

Niko nods: "*Apis florea* also covers the comb with a thick layer of worker bees (Fig. 24).

Otherwise, everything is completely different from *dorsata*. The dwarfs even stop flying when you approach. And the comb is built differently, not under the branch, but completely around it. Above the branch they construct a platform, like an American football, elongated, but very round at the top. Under the branch, the comb becomes narrow. There are located the brood, the queen and the nurse bees. The bees take off and land on the upper part. That's where the honey is stored.

A. florea has a very spectacular behavior when it rains. The bees function like shingles on a roof. The rain drops run down and the comb remains dry (Fig. 25).

During the evening review of my observations, I realized that the behavior of *A. florea* should be a promising research project!"

"I can't count how many research projects are buzzing around in your head," Gudrun remarks, shaking her head.

"Yes, you're right, I have to write it all down as quickly as possible. It's all so exciting! The dwarfs were a surprise on the first day of observation. Suddenly, disorder arises on the surface of the nest. The bees run around quickly, and stop after a short run. Then more than half of the bees fly off. My log book is suddenly full of yellow droplets of excrement—so the cleansing flight is similar to that of the giants. And because I was sitting close to the colony, I could even see the comb with the much larger queen. In the lower area is the brood area for drones. And under the drone cells hung twelve queen cells. They looked like acorns similar to those in our honey bees. And two days later I saw that there were significantly fewer bees.

"The comb was no longer completely covered by the bee curtain. I discovered a young queen hurrying quickly over the comb. The next day the comb had fewer bees but there was another queen who then flew away with the rest of the bees. The comb remained without one bee. I was amazed, because in our western honey bees the last daughter inherits the cavity and all the combs with honey and brood. Nevertheless, I was satisfied and surprised about the quick end to my observations. While I was still completing my protocol notes, I observed some *A. florea* workers landing on the abandoned comb. Were these bees that had missed the connection to the last swarm? After a while, the behavior of these bees was clear: They gnawed on the walls of the empty honey cells and collected the gnawed wax particles like pollen in the baskets on their hind legs. This resulted in a lively air traffic." (Fig. 26)

The famous Karl von Frisch, Professor Lindauer's teacher, to whom I sent photos of it from Pakistan for his new book, called this behavior 'moving with furniture'."

"Great, that's a real result. You have to publish it!"

"Karl von Frisch did that and credits me as the photographer."

"It was probably all a bit much for you at once. And the amoebic dysentery and the return transport."

"Yes, that's true, but the transport was unexpectedly easy this time. As soon as the mountain queens had produced brood in their new colonies, I booked the air freight to Frankfurt with the PIA, Pakistan International Airline. There was a total of eight colonies, three of them with the mountain queens. PIA was competent and flexible. The air freight number was sent by telegram to the Institute in Oberursel. I then had to wait another three days until I got the confirmation of safe arrival. And at the end, I have you and you finally have me back!"

"Yes, we missed you very much"

"Six weeks without you and the children were far too much. I will take the three of you with me in the future."

"And then all four of us come back sick? We will have to talk about this later in peace. Now you have to get completely healthy."

"I'm very serious. I would like to do research there for at least half a year, as we did in Israel. And with all of you. In Tarnab we can live somewhat like Europeans, at least I never got sick there, but got well again. It will certainly take more than half a year until I have thought this through and submitted the application. And it takes a long time again until it is approved and everything is organized. Then Anna and Fritz will be older!"

Gudrun is too tired to contradict.

In the weeks that follow, Niko has himself repeatedly examined for amoebic dysentery, always with a negative result.

Twenty-sixth chapter

Gudrun is involved in the "Kinderschule" and in the local SPD (Socialdemocratic Party Germany)

Niko's reports from Pakistan impressed Gudrun and frightened her at the same time. For her, it is clear that she never wants to travel to such a male-dominated country. She thinks longingly back to Israel and her time at the Institute. When Niko comes home in the evening, he usually talks enthusiastically about his experiments. The foreign guests he brings home rave about the spirit of optimism at the Institute. In the process,

Fig. 24. A dense envelop of worker bees covers the comb of *Apis florea*.

Fig. 25. Rain protection of *Apis florea*. Worker bees function as shingles to prevent the rain from entering the comb.

Gudrun learns that Professor Ruttner has assigned her doctoral topic about the filling of the spermatheca after the queen bee's mating flight to a new Ph.D. student named Markus. He is to continue what Gudrun has started successfully. Gudrun knows that published research results and scientific techniques are widely available. Nevertheless, Gudrun is very affected that she was not asked or at least informed by Professor Ruttner. Only when difficulties arise and many experiments fail does Professor Ruttner ask by phone if she will help Markus. Gudrun doesn't want to cancel the experiments, but

Fig. 26. *Apis florea* collecting wax from their previously abandoned comb.

it's not easy for her.

When she asks the other mothers at the edge of the sandbox what they would like to achieve besides the children, she usually hears answers that are repulsive to her. Again and again, these women assert: "I want to support my husband. He can then devote himself entirely to his profession. Behind every successful man is a great wife. You know that, Gudrun! And a good professional career expands our financial possibilities. Get out of this rented apartment, have your own villa and nice trips. I want to contribute to that." Gudrun's further question, "And what do you do when your children grow up and leave the house?" It is often answered enthusiastically as follows: "I'm looking forward to that. Then I can finally do something for myself: play tennis, yoga, read, and go to the theater. There are so many beautiful opportunities to realize one's own interests."

Gudrun doesn't want a life like that! There is certainly nothing that is more important than raising children. No doubt she would choose it again. Does motherhood necessarily mean giving up your own career? Scientific arguments for maternal renunciation of work are proclaimed by psychologists and physicians. Now that men need jobs in Germany after the war, it is emphasized how much children depend on the care of their mothers. In novels, on television and in newspaper articles, mothers are always to blame when children's development goes wrong. Then it is said that the mother has worked and therefore did not take enough care of the child, or that the child has gone astray even though the mother was at home. According to this public opinion in postwar Germany, children are not allowed to go to kindergarten until they are five years old. A look at neighboring countries, especially Scandinavia, shows Gudrun that early childhood education in kindergartens and day care can be successful.

Then Gudrun reads in the newspaper Frankfurter Rundschau that the Kronberg "Freie Kinderschule (Free Children's School)" has moved to Schwalbach. This children's school is based on the Kinderladen movement, which was founded in West Berlin in 1968 by the student action council for the liberation of women. The article also emphasizes that the main goal of the children's school is above all a "repression-free" education of the children. The children should learn to be considerate of each other and learn to live without conflict. Particularly important: Children are admitted from the age of three.

Gudrun immediately calls the telephone number provided and contacts the parents' initiative. Both, she and Niko, are invited to the parents' evening. It's four days after the phone call. A neighbor takes care of the children. When they arrive at the barracks behind the public swimming pool, the door is open. Many parents sit on children's chairs and talk animatedly. A woman greets them: "How nice that you found your way here. I am Rose and I am a member of the board of the Kinderschule. The parents' evening is about to begin. Please sit down somewhere!" As soon as Gudrun and Niko sit on a mattress lying on the floor, Rose introduces them: "These are Gudrun and Niko, they have two small children and would like to join us! I told them on the phone that we don't just take in children. We strive for an education that responds to the needs of the children. That must also apply at home."

Accustomed to speaking in front of large groups, Niko answers: "Yes, this is my wife Gudrun and I am Niko. Our daughter Anna is just under three years old, our son Fritz is one and a half. We have followed the reports of "Kinderladen" ideas and our education is also geared towards independent behavior. The children should have their own experiences and we try to explain if not everything goes as they had planned."

Peter interrupts: "It's not as easy as you imagine. It's not just about the children, we are a parents' group here that also sets itself political goals. We want to raise children to be thinking people who do not simply obey orders like many of our parents. They should learn to question everything."

Gudrun is annoyed and interrupts Peter: "I can't get involved if I'm busy with the children all day. Before we had children, I demonstrated and, for example, tried to prevent the delivery of the Bild newspaper after the murder of Benno Ohnesorg."

A general murmur ensues. Rose resolutely settles the dispute: "I suggest that Gudrun and Niko take a look at the children's group in the morning, what and how they play. They should also take part in the upcoming parents' evenings, then we will get to know them better. And now let's start talking about the children. What problems did you have here in the children's group and what did it look like at home?"

The behavior of the children is then discussed in detail by the supervisors.

"We are still missing two lunches for next week! Who can bring the food on Tuesday or Friday?" After two mothers have come forward, the evening is over.

Gudrun and Niko talk at home about concept of the "Kinderschule". Should they get involved in this group? They also think that something has to change in society. Their relatives and their teachers were influenced by the Nazi era. "At school, obedience was also the top priority. Think of the many punishments, standing in the corner, detention and the punishment work for nothing!" adds Gudrun.

"Yes, I think the idea of the children's school, that the children should learn social responsibility and political interest from an early age, is good," says Niko.

After a few parents' evenings, to which usually only one of them can go, their application for admission to the free children's school is received positively by the parents' collective. Shortly after her third birthday, Anna is admitted to the children's school. The parents' evenings are important lessons for Gudrun. Usually the fathers discuss, the mothers are rather quiet. When individual mothers try to express their opinions they are confronted with counterarguments, women are usually not experienced enough to defend their ideas as convincingly as many men. Gudrun is no different. After a year and a half with small children at home, she has forgotten how to formulate her thoughts quickly and precisely.

"And this only two years after my presentation in front of an audience of more than 1000 at the Apimondia Congress in Munich!"

Gudrun is determined to change her situation. Nothing comes from nothing. Gudrun organizes women's meetings where the silent mothers can prepare for the parent discussions. Their goal is to formulate thoughts aloud in advance so that they are not immediately outvoted by the fathers. The three women, who are actively involved in the discussion at the parents' evenings, do not take part. They are now working and have little time. The other women show no interest in having a say themselves: "My husband does it very well. He knows better than I do how to represent his point of view. It's more important that someone presents it with persuasiveness."

These evenings often end as if at the sandbox with the question: "What are you cooking tomorrow?"

Anna usually likes to go to children's school. She enjoys making fires, swimming, building birdhouses or going on trips to the fire department, workshops such as car repair or the zoo. The children's interest in their environment is stimulated and implemented in role plays, painting and handicrafts.

Inspired by the many political discussions of the parents' group, Gudrun decides to become active in local politics. "I want to be particularly committed to the situation of women and children. I have some experience in that now," she explains to Niko. She starts on a small scale and goes to citizens' and SPD meetings in the evenings. The following year, she was elected to the board of the local SPD. Together with SPD women and the Kinderschule, she organizes demonstrations on Mother's Day: "More kindergarten places!" and "All-day school now!" are the demands. First of all, it is mainly about the many mothers and single parents who have to work for financial reasons and whose children are disadvantaged due to a lack of care. Some politicians come to the demostrations and they agree.

The demonstration do not move much, many SPD comrades have internalized the traditional image of women and keep finding new reasons why nothing can be changed. Gudrun is growing frustrated by her futile efforts to make improvements here and now. In addition, there are personal disappointments. A young comrade, who represented left-wing, grassroots democratic positions, is supported by the SPD women in the board election and wins. Hardly on the board, he allies himself with the mayor and prevents the decided construction of the adventure playground in Schwalbach. Confronted by Gudrun, he admits that he was promised a position in the magistrate. Intrigues and agreements behind the backs of those affected also poison the atmosphere in the local association and lead to Gudrun questioning her commitment to the SPD in Schwalbach:

"Niko, I don't want to go on like this. Always only disappointments and failures. In addition, the many personal insults and attacks! I am unsuitable for political party work. What can I do?"

Niko remembers a saying from his father: "Many enemies, much honor. You are successful and that is why many comrades are afraid of you. They see their position endangered by you."

"Niko, always your stupid sayings. I'm tired of that. I want cooperation, trust and maybe friendship and you come to me with 'Many enemies'. It's getting harder and harder to talk to you. Are you even listening to me?"

Gudrun runs out of the living room. The door closes loudly. Niko remains affected. "Gudrun is probably right," he thinks after he has calmed down: "In the last few weeks, Gudrun has been on the road a lot and when she came home, usually late, I was too tired to listen to her seriously and talk to her."

Twenty-seventh chapter

Gudrun finds her own way

The next morning, Niko has a plan: "Gudrun, you should come back to the Bee Institute. I'll talk to Professor Ruttner, he'll be happy!"

"Niko, that's not possible! My topic ended up with this Markus. Professor Ruttner has decided that. Besides, the two of us at the Institute? Anna is in the children's school. Tell me, where should Fritz go?"

Niko does not give up: "Gudrun, I will find a good solution to all problems. Come back to the Institute."

"Niko, finally understood. I don't want to return to the Institute. I don't want you to look for solutions for me. I want my own way!" Gudrun gets loud, she has tears in her eyes. Niko reacts dismayed. He has to go, to a meeting at the university: "We'll continue talking tonight. Bye."

In the evening, Gudrun rejects any conversation about Niko's proposals.

Fig. 27. Free Children's School in the City Schwalbach 1973.

"Then just not," he says loudly and angrily. After a short pause, he continues more calmly and conciliatory: "I will wait and help whenever you change your mind."

Gudrun is sad and relieved at the same time. She can understand Niko's anger. She is sure: Only without Niko's interference, she will find her way back to her own feet.

Gudrun wants to find a job as a biologist. Her options are limited. She can work one day a week, or better only half a day. How is that supposed to work? Meanwhile, at a party of her friend Bärbel, Niko stays at home looking after the children, she happens to meet Eva and Heiner, former biology students whom she has known since she moved to Frankfurt for her studies.

"Long time no see. How are you doing? What do you do?"

"Hello Gudrun! Nice to see you!" Heiner greets her with a short hug. "Where did you leave Niko? It's the first time I've met you without him. My big chance?" Heiner grins broadly. "Heiner is still the same one-liner," Eva hisses. "You know him. Heiner and I have completed our studies. I work in Bad Soden and Heiner is at the gymnasium (high school) in Oberursel. We have become biology teachers."

"What is your son doing? He should be two years old now?"

"Jochen has just celebrated his third birthday. He is our great joy," Eva beams.

"And how do you reconcile work and child?" asks Gudrun, looking curiously at Eva.

"Nothing could be easier. My mother is happy about her grandson Jochen, we all live together in her house. I have a three-quarters job. There is enough time for my son. However, I can't say the same about Heiner. He tinkers with his cars in his free time and doesn't take care of his family."

"Eva, leave Gudrun alone with your accusations. I have a full-time job and have recently started working at the seminary for trainee training in Frankfurt. Unfortunately, there is little time. Gudrun what are you doing?", Heiner continues. "The other day I heard that Niko has started his work towards the qualification to teach at a university. And what do you do? Still queens and honey bee reproduction? Are you still at the Institute in Oberursel?"

After this direct question, Gudrun has to swallow first. She answers haltingly and in a low voice: "No, I'm no longer at the Institute. My topic has been taken over by a new doctoral student. As you know, we have two children, and that's my job at the moment. Niko arrives late and goes to the Institute early. Children and household depend on me." Gudrun struggles to keep her composure. "Gudrun, you can change that. Why don't you come to us at the Gymnasium in Oberursel. We are desperately looking for biology teachers. Don't you want to take on a teaching assignment? Maybe two courses as an introduction?"

Heiner sounds convincing. Gudrun is surprised: "How is that supposed to work? I never studied teaching. What are the topics in class?"

Heiner is an excellent salesman and has realized that Gudrun is seriously considering his proposal. "Gudrun, you can do that with your left hand. There are also the specialist conferences. All biology teachers of the upper classes discuss their topics and the teaching problems. In the beginning, this could be an important help for you. And for my colleagues, a thoroughbred scientist with a doctorate like you is definitely a great asset." Heiner is enthusiastic about his proposal.

"Now do it halfway, Heiner," Eva objects. "Such a biology course requires thorough preparation and a high level of professional competence. Gudrun is quite right with her objections. Besides, this is a party and not a conference. Come on, I want to dance!" Eva takes Heiner by the hand and pulls him onto the dance floor. Gudrun pours herself a large glass of whisky. Unfortunately, there is no bourbon, only Scotch. To school? Should her father ultimately be proven right? Gudrun needs a big sip. Even before she can finish her drink, Bärbel's husband Rolf comes and asks her to dance. "Gudrun, first a dance and then your drink?" "Don't be angry Rolf, first this sip and then a dance!" "One moment. I need a glass as well. First toast, then drink, talk and finally dance?" Before Gudrun can answer, Rolf is gone. He comes back with a well-filled glass. "Cognac?" "Yes, and you?"

"Whisky. Cheers! Here's to your well-being and this beautiful party!"

They toast. After the second sip, Rolf asks: "Tell me Gudrun, what's going on? You look pretty battered." "That's not true. I am a happy mother and housewife–like many millions of happy women, some of whom I meet every day at the edge of the sandbox on the children's playground." "Hmm, so bad? Dissatisfied?"

"Rolf, I need work. And at your wonderful party, a few minutes ago, I got an offer. Biology lessons at a school. I never wanted to be a teacher." Gudrun moans and takes another big sip. "Well, the harsh reality. Only this chance and not your desired concert! I understand. You don't want that? What is the law of biology that I learned in my school days: Eat or die a bird!"

Rolf's glass empties. Gudrun lifts hers: "Cheers, Rolf, I understand, you mean I should do it, even though I don't want to and can't. Yes, I'm particularly good at things I don't want. I will follow your advice. Do you really want to dance with me?"

Rolf wants to get up. That doesn't work. He falls back into the chair: "I think I have to postpone this dance to our next party."

"No problem, I don't feel like dancing either. I'll go home. Niko is certainly waiting. And thank you very much, Rolf. You have helped me!"

Gudrun bends down, a short hug and then walks home with slightly buoyant steps.

Niko lies in bed and reads. "Gudrun, how was it? A good party?"

"You're still awake? Yes, maybe too much whisky, but otherwise successful."

Gudrun yawns: "I can tell you the rest tomorrow morning."

She disappears into the bathroom, leaving a completely surprised Niko behind. The next morning, Gudrun has a headache and asks Niko for one of his painkillers. Niko gives her a Gelonida and asks: "Gudrun, you wanted to talk about yesterday's party. What was going on?" "I met Eva and Heiner. They send their greetings to you. Both are now teachers. Heiner is at the Gymnasium in Oberursel. Then I drank a large glass of whisky with Rolf. There was only Scotch. Hence my hangover! Otherwise, it was quite nice!"

Niko suspects that Gudrun is not telling him "everything". It's half past seven. He urgently needs to go. Martin Dautz is waiting for him at the Institute.

In the next few days, Gudrun calls Heiner several times. Then there is an appointment with the head of the Gymnasium in Oberursel and just two weeks after the party, Gudrun holds a teaching assignment for grades 11 and 12 in her hands.

Of course, Gudrun's activities did not escape Niko. He has not forgotten Gudrun's rejection of his offer of help. So, he doesn't ask.

By chance, he meets Rolf on Saturday while shopping: "Hello Niko, did Gudrun arrive home safely after our party? I had a terrible hangover. Your wife is really good at drinking." "Yes, Gudrun can take a lot. Much more than me. Then it must have been boozy and cheerful with you the other day?"

"Well, Gudrun was in a pretty bad mood. Tell me, what has she decided now, will she go to school and teach biology? Probably not an easy decision. Probably the only right one. Gudrun has to get out. The sooner, the better! I strongly advised her to do so, as far as I can remember."

Niko doesn't want to admit that Gudrun didn't inform him. That's why he is quite general: "I think the negotiations are still ongoing. I also hope she will accept the offer. Rolf, I have to go. I want to look for chestnuts with the children. Goodbye! Greetings to Bärbel and come by!"

Niko has difficulty accepting Gudrun's behavior. The conversation with Rolf has further unsettled him. Gudrun doesn't talk to him about her problems, but to Rolf. Niko sees this as an insult and rejection. It's not about him, he thinks further. He kept everything, research, the Institute, the university, many trips and congresses. Gudrun has given up all that and feels the loss painfully. In this difficult situation, Gudrun does not need an offended husband, but support and help in her departure into a professional future. Niko decides to keep a low profile and wait until Gudrun talks to him about her plans.

Niko goes into the forest with the children. Anna and Fritz's chestnut basket is soon well filled, to their delight. "What do we do with the chestnuts?" asks Anna.

"We'll bury them," Fritz answers. "Every chestnut becomes a new chestnut tree."

"Or we take the chestnuts home and make a zoo with chestnut animals out of them. I'm sure Gudrun knows how to do it," Niko suggests. "First we'll bury three of them. That gives three new trees." Fritz is starting to dig. On the way back home, some acorns and spruce cones are collected. Gudrun admires the treasures they brought and fetches matches and toothpicks: "The chestnut animals also need legs and tails." Before dinner, a magnificent zoo is created. Fritz has finished three chestnut bears. Anna has two elephants and Niko, much to the delight of his children, a fat chestnut bee. Gudrun has made a giraffe, two lions and a parrot. Then they eat and the children are supposed to go to bed. Anna gets the elephants and the giraffe set up in front of her bed. Fritz his bears, the lions and the parrot. Niko's chestnut bee remains. "The bee is coming to you!" the children proclaim.

When Niko comes back into the living room, there is a bottle of red wine and two glasses on the table. "Niko, I want to talk to you." Gudrun seems tense.

"The chestnut zoo was fun for the children. And us too? Let's toast, Gudrun. To the chestnut animals!" Gudrun and Niko take a sip. "Kaiserstuhl Pinot Noir, how good! Gudrun, what's wrong?" "I was given a teaching assignment at the school in Oberursel for two courses. Grade 11 and Grade 12. Both courses are on Thursday morning. Can you look after Fritz during this time?" Gudrun looks at Niko. "I can only sign when I have your acceptance." "I can arrange that. Yes, I'll take care of Fritz." Gudrun is surprised by his unconditional acceptance. She raises her glass: "To Niko, my very best husband!" "Here's to Gudrun, the very best wife and mother!" Both are happy that the status quo is coming to an end. They are not afraid of the concrete challenges that now lie ahead of them. After all, they have mastered so many difficulties together.

Gudrun familiarizes herself with the topic of "sociobiology" in the evenings. With her experience with honey bees, this is not difficult for her. She involves Niko in the course preparations, which leads to the reactivation of the proven and successful co-operation. Gudrun creates extensive documents and concepts for her lessons. At school, however, things don't go as she expected after the conversations with Heiner. The promised preliminary discussions with other biology teachers do not take place. There are also no meetings with discussions about learning objectives and problems. In order to gain experience, Gudrun attends the lessons of two colleagues several times. For this, Niko has to take extra time for Fritz. The lessons of these biology teachers are just as boring as they were during her own school days. As a student, Gudrun didn't like to cram by heart, and as a teacher she doesn't want to demand that either. It is intended to help students understand biological relationships and modern sociobiological approaches and hypotheses. Such topics are not found in the official curriculum. Teaching that is mainly reduced to multiple choice questions, on the other hand, contradicts Gudrun's intention. Finally, it becomes clear to her that she does not want to enter the school teaching business under any circumstances. After this winter half-year, Gudrun decided to leave the Oberursel high school. She rejects several offers and requests from the school management to continue teaching next summer.

Twenty-eighth chapter

Gudrun returns to bee research

Confronted with biology lessons at school, Gudrun begins to consider whether she should return to the Bee Institute after all. What can she do there? Her former topic is taken. One evening, Niko brings a publication about butterflies, in which it is described that the males of this species transfer chemical substances during mating, which cause the females to start laying eggs quickly. Gudrun reads the publication and is enthusiastic: "Could that work similarly with honey bees? We know that queens lay eggs within the two days after the nuptial flight, but after instrumental insemination, in which only sperm is injected, it takes weeks." "Yes, I thought so too," comments Niko.

Gudrun becomes excited: "Imagine the drone has such a substance and you could identify such a substance and then synthesize it. In instrumental insemination, this substance could be added. That would save the second CO_2 anesthesia the next day." And she is thinking about it: "I can master the techniques for the experiments! With my surgical technique, I can inject drone glands or their secretions into the body cavity. And I can also do instrumental insemination. I could add glandular extracts to the sperm during insemination."

Her head is swirling. What is transferred to the queen during mating? Does the mucus secretion or the orange secretion from the mating organ play a role? The longer Gudrun thinks about it, the clearer it becomes to her that this question offers an extremely promising new approach to research. Niko beams: "Finally the old Gudrun again. This is internationally "top priority." The instrumental insemination of queen bees is becoming more and more important worldwide."

"On this topic, I would be independent of the time of day. I could carry out both insemination and surgery in the morning and do not have to be with the bees in the afternoon at the peak flight time, as in the last phase of my doctoral thesis. Fritz will be two years old next spring. Then I'll look for a friend who will look after Fritz in the mornings during the next bee season. Do you think Inge would do that? You know her from the parents' evenings in the Kinderschule. Isn't she a great woman? Instead of cleaning telephone booths as at the moment, childcare is certainly a better job."

Since Gudrun works part-time, she can't go to the Institute with Niko. Niko buys a scooter. He does not find a Heinkel, which he would like so much. There is a well-preserved Vespa in the neighborhood. The VW Beetle remains for Gudrun and the children.

At noon, Gudrun picks up Fritz from her friend and Anna also comes home after lunch. She often brings other children from the children's school with her. Gudrun is happy about it and plays along or helps if the children want it. Fritz can also take part at the age of two. Gudrun enjoys the lively crowd of children. "Children plus bee research! What a great happiness!" thinks Gudrun full of satisfaction.

Professor Ruttner is enthusiastic about Gudrun's new research project. To practice the technique, Gudrun is the first to receive queens at the Institute in May. Despite the long break, the surgical technique does not cause Gudrun any difficulties. The introduction of glandular material into the sting chamber or body cavity is also successful. First, the substances that make up the mating sign are tested. After it is certain that the queens survive for several days, 20 queens are treated in small bee colonies, so-called "Einwabenkästen" (one comb mating boxes).

The queens survive, but like untreated queens they don't start laying eggs until four weeks after the treatments, regardless of what secretion they received. According to the report, the drones' secretions do not contain any substances that stimulate the queen to lay eggs quickly. After the bee season Gudrun will have time until next spring to think about it. However, Gudrun wants to clarify at which part of the mating process that egg laying is triggered. She plans to examine the individual sections of the mating behavior separately from each other. Professor Ruttner is enthusiastic and applies for a scholarship for Gudrun from the DFG (German Research Foundation).

The scholarship is approved in April. Fritz is admitted to the Kinderschule at the age of just three. Anna takes care of her brother, so the settling-in process goes quickly. Gudrun's experiments begin in mid-April. Professor Ruttner thinks that the flight may cause an increase in CO_2 in the queen, just like with CO_2 anesthesia during instrumental insemination. Maybe that's the trigger. Accordingly, Gudrun has to separate the flight of the queens from the mating process. Where can queens fly without being met by drones? A North Sea island "Norderney" is an ideal location for this experiment. Before the official commissioning of the "mating station", there are no bee colonies on Norderney and therefore no drones. Gudrun calls her mother in Wilhelmshaven: "Mom, I've now received a research scholarship. For this I have to do experiments on Norderney for a week, Niko will help me. Could you take the children for a week?"

The answer from Gudrun's mother is positive: "Of course, the children are older and I don't need to wash diapers anymore. That makes me happy. Maybe Lars, Hilde's third son, can also come to us, he is only a little older than Anna. All three of them can play well together!"

"Oh Mom, that's a great help! You don't even know how much this helps me with my professional re-entry!"

Gudrun and Niko drive with the children and Gudrun's bee colonies from Oberursel to Wilhelmshaven. Little cousin Lars is waiting: "Come quickly, grandma has a lot of new Lego bricks!" Immediately, the three children disappear.

Early the next morning, Gudrun and Niko reach the ferry to Norderney. After the crossing, it is still cool and the 30 test colonies are quickly set up. From the next day, the queens fly out several times for more than 15 minutes. One queen searches for drones on 20 flights. After six days on Norderney, the colonies are closed at night and packed for return transport. The next morning, the children are picked up at Gudrun's parents and they go back home. Niko drives to the Institute at night and sets up the colonies. Now it is a matter of waiting. Gudrun checks the colonies every morning and looks for eggs in the combs. In vain: no eggs, not a single one of these queens starts laying: Flying is not sufficient for the start of the egg laying! Gudrun can thus rule out three factors as triggers: first, the sperm, second, the secretions of the male glands, and third, the flights as such. What now?

Every summer, there are many guests are at the Institute. They also like to sit comfortably at Niko and Gudrun's home. Tom and his friend Liz from the USA rave about the old castles in Germany and the many knightly sagas they have heard. Gudrun, who is pondering what to do next, is overtaken by a flash of inspiration: "Many of the knights had to leave their young pretty wives alone in the castle. And what did they invent so that the beautiful women would remain faithful to them?"

Tom and Liz laugh: "That's what they explained to us: A metal chastity belt. They took the key with them on their war trips." "Bingo! I need something like that for the queens, too!" What can a chastity belt for queen bees look like? The suggestions, inspired by wine, become more and more fantastic, from gold rings to small screw clamps to chain mail. "I can't put metal parts on the queens, they're much too heavy. Maybe I could try some kind of stocking?" Gudrun's thoughts keep turning: How to prevent queens that encounter drones on their mating flights from being mated?

The next morning, Gudrun tries out various "chastity belts". A small piece of thin elastic nylon tissue that is pulled over the abdominal opening of the queen bee and glued together proves to be the most favorable. To Gudrun's great joy, the queens fly out to mate despite the chastity belt. More than half of these queens start laying eggs within a week. All were unfertilized drone eggs! Was there more contact between the queen and the drone than intended? A hole or too elastic, not so chaste chastity belt? Or is the trigger for egg-laying perceived via the open sting chamber? The mating season is over again. It will take a while before Gudrun can find an answer to these questions.

Gudrun's new experiments have consequences for the children. Queen bees only fly for mating when the sun is shining, more than 23°C and (unfortunately!) only between 2 and 5 p.m. As soon as the weather is good, Gudrun has to watch her queens and cannot go to the swimming pool with her children. She has to find other mothers to look after Anna and Fritz in good weather. This only applies to May, June and July. In August, Gudrun can return the favor and usually goes swimming with a large group of children.

Twenty-ninth chapter

Niko fetches Stefan from Darmstadt

In Pakistan, Niko realized that the defense behavior of Asian honey bees is more differentiated and complicated than the colony defense of the European *Apis mellifera*. He is sure that sounds, vibrations and visual stimuli play an important role in the enemy recognition of Asian bees. However, Niko considers his own knowledge of physics to be "insufficient". He needs a specialized cooperation partner who, in addition to the appropriate expertise, also has the necessary technical equipment. During the evening conversation with Gudrun, Professor Hubert Markl comes to mind. As an assistant to

Professor Lindauer, he published several very outstanding papers on vibration perception. His spectacular discovery is unforgettable: leafcutter ants buried in the ground emit vibration signals and are thus discovered and dug up by their sisters. Markl recently accepted a Professorship in zoology at the Technical University of Darmstadt. Niko talks to Professor Ruttner about his cooperation plans:

"Of course, I will help whenever necessary," is Professor Ruttner's answer.

Professor Markl quickly answers Niko's inquiry. He finds the planned trials interesting and promising. Unfortunately, he himself does not have time and suggests that Niko should contact one of his doctoral students, Stefan Fuchs who works on knocking signals in horse ants and is exactly the right person for the experiments outlined. Stefan Fuchs is not convinced. He will talk to him. At the end, Niko had to try to win Stefan Fuchs personally for this cooperation.

Getting Stefan Fuchs on the phone was not easy. Only after several unsuccessful calls can Niko reach him. Niko's offer to come to Darmstadt to talk to him about the planned experiments is accepted after some hesitation. Then Stefan seems to have doubts:

"If I think about it, I'd better come to Oberursel. I could look at the Asian bees right away and assess whether I can help."

Niko happily agrees to this suggestion:

"When do you want to come?"

That is probably not so easy. Stefan agrees to call back the next day and then make an appointment. Niko is annoyed and tells Gudrun about this unnecessary delay. Unexpectedly Stefan calls in the late afternoon. He wants to come next Monday: "Can I see these Asian bees then? I would also like to observe their defensive behavior. How do you know that the behavior described is for defense?"

"I have an *Apis cerana* colony in an observation box in a heated room. There I can demonstrate the two essential behaviors at any time. And at any time speaks for defense, doesn't it?"

"Yes, however..."

Niko interrupts: "Couldn't we discuss this better on Monday?"

On Monday afternoon, a dented yellow VW Beetle—probably retired by the post office and bought cheaply at auction—stops in front of the Institute. A slender, bearded student with round nickel glasses gets out and comes to the entrance. Niko goes to meet him:

"I'm Niko. Welcome to the Bee Institute!"

"Yes, I'm Stefan. We had been on the phone? And now I'm here!"

They talk about the program. Stefan wants to return to Darmstadt around four o'clock. He doesn't want to miss a seminar. First there is a cup of coffee for Stefan and Niko. With the cup in their hands, they go upstairs to the bees. In the room with the observation hive, only a lamp with a red light is burning, which the bees cannot perceive. Stefan looks closely and is disappointed:

"This *Apis cerana* hardly differs from our western bee! You say that the behavior should be different in principle?"

Instead of an answer, Niko knocks lightly on the beehive with his coffee cup. The bees react immediately and respond with a loud and clear hiss. Stefan looks surprised:

"Niko, do these bees synchronize with each other? Or maybe everyone reacts at the same time, independently of each other? We have to set a locally clearly defined stimulus and then see if the hissing goes beyond!"

"A good suggestion! This is difficult to achieve by knocking on the beehive. Can you record the hissing sound? Do you have such microphones in Darmstadt? We could play back the hiss. Maybe that's a good trigger?" Stefan looks at him doubtfully:

"As far as I know, bees and ants can't hear. Their sound perception runs through vibrations of the substrate. The hissing sound from our loudspeaker should make the comb vibrate."

Stefan takes a breath to elaborate on his thoughts.

"Dear Stefan, it's going much too fast for me," Niko interrupts. "Shouldn't we first discuss only the first experimental step and then arrange the date for it?"

Three days later, Stefan and Niko are sitting in front of the observation hive again. The front glass has been removed. The bees remain calm and besiege the combs with the brood. Following Stefan's instructions, Niko carefully attaches a microphone which is connected to a tape recorder to the comb. This is how the first sonograms of the sibilant sound are created, which can be analyzed. Their experiments to trigger the hissing behavior with the speaker are successful. Frequencies around 50 hertz, i.e. low humming tones, are most effective. At tones around 300 hertz, there is no reaction even at high intensities. Niko is thrilled with these clear results. It has become late. Niko had promised Gudrun to bring the children to bed with her. Because they want to continue tomorrow morning, Niko invites Stefan. He is a little hesitant, and then comes along. Gudrun is happy to see Stefan and the children are also enthusiastic. A visit in the evening, they know from experience, it means staying up longer. When Anna and Fritz are in bed, Stefan, which is her wish, has to say "good night" to them.

In the meantime, Gudrun fetches a bottle of red wine and asks Stefan about the current experiments. He reports on technical problems and puts the results into perspective:

"We are still far from knowing the decisive connections."

"Niko was thrilled this afternoon when we talked on the phone and you mean that doesn't mean anything?"

Stefan reacts irritated: "Niko is right. We have made good progress. There are still so many details missing. We have to be careful. The real difficulties are still ahead of us!"

"Dear Stefan, we have seen that a bee colony collectively produces a hissing sound for defense. That's great!" interjects Niko.

"Great?" asks Gudrun. "That's typical for Niko. He often goes a step too far. I agree with you, Stefan. One must first prove that there is communication between the individual bees before one can speak of a communal behavior. Find out how the hissing signal travels from one bee to the next."

Stefan is happy about Gudrun's support. Niko says: "Two against one, that's not fair. We will soon learn how the signal is passed on within the colony. Tomorrow, we want to check whether the bees perceive the sound through the air or through the vibration of the comb. Stefan has an idea. It's late. The children wake up early. We should go to sleep."

The next morning, a concrete block weighing more than 150 kilograms has to be heaved up the stairs into the observation room. Strong students and beekeepers are in demand. After a short time, they have to take a break as they continue to maneuver the block up the narrow wooden spiral staircase. The helpers indignantly ask whether a lighter object would not also meet the scientific standards.

"Bees are deaf," explains Niko. "They can't perceive airborne noise. They react very sensitively to the vibrations of the substrate and the combs. Yesterday the bees reacted to very loud deep noises, now we have to find out whether they heard the airborne sound or rather perceived the vibrations of the comb. This concrete block is so massive and heavy that it is not made to vibrate by the loudspeakers. If the bees sit on the concrete block and react to sound, we would have an indication that bees are not deaf and may be able to hear a little after all."

The initial protest about the heavy load gives way to a lively discussion about the deafness of bees. However, they only manage the last stage into the observation room by cursing loudly.

Stefan and Niko glue a small wire cage with the queen and some bees in the middle of the concrete. They carefully sweep the bees from the combs onto the cloth in front of the block. After the bees have gathered on the concrete block around the queen, Niko carefully blows on the bees. To their delight, the answer is a loud hissing reaction. Then the speakers are used. They start with relatively quiet tones. They do not react to the sound system. Blowing, on the other hand, continues to be answered immediately with a hissing reaction. They gradually increase the volume to over 120 dB, a deafening sound. The students in the lower rooms are complaining, but the bees are not responding. Sound that does not cause vibration is therefore not a trigger for the hissing behavior.

Thirtieth chapter

Niko and Stefan hiss at bears

In the following weeks, the analysis of the hissing behavior of *Apis cerana* progresses well. The triggering signal is transmitted from bee to bee with the wings. Furthermore, the hissing behavior causes the flight activity of the colony to be reduced. For Niko, the biological situation is plausible: In his opinion, hissing noises from a dark hollow tree are "snake mimicry". They deter bears and other predators. Stefan, on the other hand, is of the opinion that such speculation is too far-reaching: "We need field studies to understand why and against which predators the hissing of the bee colony is used," Stefan objects.

Waiting for later field investigations is not Niko's way. He asks for Asian bears at Frankfurt Zoo. "Yes, there are two Asian Malay bears here. Whether you are allowed to carry out experiments must be decided by the responsible research assistant. Can you submit a written application?" "No, not an application. Could you please tell me the name and extension of the chief bear caretaker?" When Niko hears the name Dr. Caroline Liefke, he is relieved. Caroline was in his internship as a student. When he calls, she asks him what he is up to and whether the bears will be harmed or pained.

"No, we want to train the bears. No punishment, only reward! In a beehive lies a piece of comb. The bear learns to knock over the box to reach the comb. In the experiment, next to the comb, there is our loudspeaker, which transmits the *Apis cerana* sounds at the first touch of the bear."

Caroline likes that, she would like to help with the experiments. Stefan criticizes this initiative: "With two imprisoned Malay bears, you can't prove anything. Such an experiment has no any scientific significance." "You're right, Stefan. For me, however, bee research is also a personal matter. Even if a scientific journal were to reject the report on the experiment—which I doubt—the gain in knowledge is still worthwhile for both of us. Maybe we will find many Malay bears in the wild later?"

Stefan laughs: "Niko, you're dreaming! Of course, I'm at the Zoo and look forward to seeing the bears. Only bees are slowly getting bored anyway!"

At the Zoo they meet Caroline. She goes with them to the barred interiors. As agreed, both Malay bears "wait" in individual cages (Fig. 28). Access to the training cage can be opened via a sliding door. They set up the wooden box with the piece of comb close to the grid, then Caroline opens the sliding door. The bear does not move. Only after a push with a pole does he trot into the test room and settle down in a corner. Caroline suspects that this animal can no longer be trained after more than twelve years of captivity. The other Malaysian bear is a young female that the German ambassador to Malaysia gave to the Zoo a year earlier. This bear immediately runs into the test room, sniffs the entrance hole of the box and pushes the box violently with its paw. The box falls over and releases the comb, which the bear eats with obvious pleasure. Caroline, Stefan and Niko watch enthusiastically. The bear has passed the test.

"Now the experiment can begin," Niko notes. "First, the bear has to go back to its cage." It takes a while for the bear to lick up the honey that has dripped down on the floor. Finally, she goes back to her cage. They close the intermediate door and bring the beehive back into position. This time there is the loudspeaker next to the comb, which is connected to a tape. With a ripcord, the speaker can be pulled through the grille. "The bear must not injure itself on the loudspeaker," says Caroline.

Stefan laughs: "And I have to save the loudspeaker from this bear's paws. The speaker is expensive and belongs to the university!"

Caroline opens the door to the test room a second time and the bear runs purposefully towards the box with the comb. As soon as she has her muzzle at the entrance hole, the hissing sound is released. The bear retreats and runs around for about two minutes without a break. Then she carefully approaches the beehive again. Her reaction to the second hissing sound is even more violent: the bear shrinks back and growls three times very loudly. Then she hurries to the intermediate door and scratches at it. It stands on its hind legs and shows stereotypical movement patterns.

The three scientists discuss: "How should we continue? Do we want to wait any longer?" Caroline becomes afraid for her bear: "The poor animal is quite distraught. I want to stop the experiment now!"

With a two-thirds majority, Stefan is against, it is decided to remove the loudspeaker now and end the experiment. The bear calms down and approaches the box for the third time. Undisturbed by the hissing, the poor animal knocks over the box and eats the well-deserved honey reward. "This is a wonderful demonstration. The *Apis cerana* colony can obviously chase away a bear with their hissing."

Stefan remains critical. He asks Caroline: "Do you have any bears that are not found in Asia?" Caroline recommends a Kodiak bear: "It's active and certainly learns quickly!" Compared to the Malay bears, the Kodiak bear is a real giant.

"A good twice as big as the Malays!" is Caroline's answer to Stefan's question.

It takes some time for them to install it in the neighboring cage and the box with honey back in the test room. Caroline opens the door again and the giant bear trots into the test room. He also walks towards the box, sniffs the entrance hole. He reaches into the entrance hole with his claws and lifts the box up slowly, almost carefully. Then the box is pushed backwards with momentum and the comb is eaten immediately. With another piece of comb, they lure the Kodiak bear back into its cage. The box, now with a crack in the front board, is set up again at the grid with comb and loudspeaker. Stefan stands at the tape. Niko holds the lifeline for the loudspeaker in his hand and on "Go" Caroline opens the door. The bear moves leisurely into the test room, sits upright and licks its front paws. Then he trots to the box. Stefan triggers the hissing. The Kodiak bear hits the box with his right paw at lightning speed, which completely collapses. At the last moment, Niko manages to pull the loudspeaker through the grille and bring it to safety. After the massive paw blow, the bear backs away a little and growls loudly. Then his muzzle slowly approaches the smashed box. He seems to be sniffing. Only after another six minutes does the bear finally lick the honey and the comb with remains of wood splinters quite hesitantly and carefully. The Kodiak's violent reaction to the sibilant sound has frightened the scientists. Caroline is relieved that the experiments are over and Niko is convinced that sibilants to ward off bears would be of no use to a bee colony on the Kodiak Islands.

Thirty-first chapter
Gudrun and Niko attend a congress in London and Niko becomes a lecturer

The Congress of the International Union for the Study of Social Insects, IUSSI for short, takes place in London in 1973. Professor Ruttner calls Niko into his room:

"You know that you absolutely have to register a lecture there, right? This is a stage, Niko, on which a performance is worthwhile. Everyone who has rank and name in research on social insects will listen there."

What will Gudrun say? Alone again in Schwalbach with two children, while he meets important colleagues and friends in London? When Niko brings up the matter at home, Gudrun has a surprise in store: "I'm going with you! On the way to Ostend we can make a detour via Wilhelmshaven. Just yesterday my mother complained that she sees her grandchildren so rarely. I'm really looking forward to the Congress and London." Gudrun hurries to the phone and talks cheerfully and insistently to her mother. Then she beams: "Yes, that's okay. Mom is looking forward to the children. I just have to give her the exact dates tomorrow. She wants to gently prepare dad for the attack of the grandchildren." To Gudrun's great relief, the children like to stay with their grandparents. Gudrun and Niko are looking forward to the "child-free" time. Then, on the way to the ferry, they only talk about Anna and Fritz. "What has become of us?" asks Niko. "Of course, a completely normal family! What were you thinking?" answers Gudrun. On the English Channel, a different topic is taking center stage. They are standing at the bar and toasting: "Here's to a week without children! The second time in three years!" Suddenly, they almost spill their whisky. They feel strange large gentle swings that end with an abrupt impact. Sailors and technicians run back and forth excitedly. Even the cooks in full gear hurry to the railing. Then comes the announcement: The ferry has hit a sandbank and is stuck. The captain expects that the ship will be released under its own power when the tide sets in. The first passengers hurry on deck with pale faces, the swinging movement of the ship is hard to bear. A family with four small seasick children does not reach the upper deck in time. They help: Gudrun carries the little girl, who is shaken by severe nausea, upstairs. Niko gets a mop bucket and a squeegee and removes the mishap. Later, Gudrun and Niko are back at the ship's bar with a small

Fig. 28. Malay bear (*Helarctos malayanus*) at Frankfurt Zoo.

group of seaworthy people. You hear stories about seafaring and what can happen in the process. Gudrun smiles. As a former child from the North Sea coast, she is probably more familiar with fantastic tales of the sea than Niko, who wonders about the imagination of the otherwise rather boring North Germans. "I'm sure no one will believe us, it's all just "Seegarn" (sailors' stories)."

A delay of four hours in Dover is even favorable for them. Gudrun notes happily: "Instead of arriving at three o'clock in the morning, we don't arrive until around 7 o'clock. Instead of sitting on the cold street, we can sit comfortably in the warmth here!" So, they enjoy their time in the ship's bar drinking Coke, in 4 hours it will be necessary to drive a car with left-hand traffic.

In London, they have booked the cheapest accommodation. The accommodation at Queen's College is spartan. The historic multi-bed dormitories are separated by gender and date back to past centuries. The Congress is worth the inconvenience. Many exciting lectures! They feel sorry for their French colleagues, who, as prescribed by their institutions, have to give lectures in their native language. As a result, more than half of the audience often leaves the lecture hall and only comes back when a lecture follows in English. The situation is even more drastic for contributions in German. Niko is pleased that he has reworked his lecture in English following Gudrun's advice and against Professor Ruttner's recommendation. A full auditorium has a lively discussion about the demonstration experiments with the bears. In the evening, Gudrun and Niko sit together with English and French colleagues. Some colleagues make critical remarks about the dominance of English. Is it only about the best scientific communication? Some French people declare: "If you continue like this, you will lose German as an international scientific communication medium. The exchange from perfect, scientifically precise German to primitive English means an irretrievable loss." Niko's answer may be a bit contraversal: "The world is supposed to recover from the German dominance? We know that. The dice has been cast. English has become the language for international scientific communication. Regardless of what we Germans or our French neighbors decide. The sooner and the better we all communicate in English, the faster we will connect with the international scientific elite."

Some German and French colleagues protest vehemently!

Gudrun positions herself more diplomatically: "Maybe it's not about questions of principle now, as Niko thinks, but simply about speaking English in London and French at the next lecture in Paris?"

Many agree with this, and the English and Americans also nod eagerly.

"I lack the belief that Ammies or Tommies will lecture in a foreign language. Even if they wanted to, they can only speak English," Niko whispers in German in Gudrun's ear.

After the appointment of Professor Ludwig von Friedeburg (SPD) as Hesse's Minister of Education, there are far-reaching changes at the universities. He wants to limit the power of professors and strengthen the non-tenured faculty. Thus, all university assistants are appointed lecturers. Lecturers of a "new kind", as some professors now disparagingly call them. As a lecturer, Niko can register and carry out internships independently, officially the disposal of the Institute and all resources remains with the Professor alone. In practice, therefore, not much will change, only the entry by name in the course catalogue. Everything else must be consensual as before, i.e. dependent on Professor Ruttner.

During his internship in Oberursel, Niko gained a good relationship with some interested students who asked him about topics for their diploma thesis under his supervision. This brings Niko into a possible conflict with Professor Ruttner. Niko's suggestion, however, to appoint him as an official second supervisor meets with Professor Ruttner's approval: "We can use every good student at the Bee Institute. The more, the better! And if the main care is left to you, Niko, that's just fine with me. I don't know how I'm going to manage everything that comes my way every day."

Two students are registering their diploma theses: Jochen Grau wants to carry out a comparative study on the alarm pheromones and sting morphology of honey bee species. Ralph Bauer is responsible for investigations into the warming of the brood.

Thirty-second chapter

Niko changes his travel plans and Gudrun arrives in Sri Lanka with the children

Finally, Niko receives the letter from the DFG with the long-awaited approval of his application for his research stay in Tarnab. Niko immediately develops a plan on how to drive to Pakistan by car and with large equipment and work there for a longer period of time. In the evening, Niko and Gudrun sit together. Unlike the enthusiastic Niko, Gudrun does not believe that she can go with the children:

"Even if they are now three and four years old. A trip via Turkey, Iran and Afghanistan takes ten days rather than a week. The children can't do that yet. And we can't spend the night with them in some dive bar like we used to."

"We can certainly do that. We have so many colleagues in Jugoslav, Turkey and Iran. We can stop at home with them."

"Oh, Niko, you're crazy again! Only to Serbia we sit in the car for 20 hours. With small children, we often have to take long breaks. No, you have to do it alone or with your students! Also, it is difficult for me to work as a woman in Islamic Pakistan, especially in rural areas like Tarnab. Fully veiled at 45°C."

Niko is disappointed: "If necessary, you could fly there with the children."

Gudrun shakes her head in despair: "Everything is always easy for you. And the children and I will have to pay for the consequences!"

In the meantime, a civil war is spreading in Pakistan, which rages in 1973 between the feudal rulers of the provinces, the ethnic groups and the government of President Bhutto. Thousands of civilian casualties are reported. This makes all plans and discussions about how to research honey bees in Pakistan obsolete. The responsible clerk, Dr. John from the DFG, calls: "The DFG will not fund a research program in such a politically uncertain country. If you want to use the approved funds, you should look for alternative locations. A corresponding written application for rezoning must be submitted to me as soon as possible." She further promises that she will proceed quickly and unbureaucratically and will only ask one of the two experts for a statement. "If you quickly find an Institute in another Southeast Asian country, you will know after three weeks!"

An alternative location for the planned trials? How is Niko supposed to do that? It should start in four months. Then Niko remembers the classic work from Sri Lanka from 1956, in which Professor Lindauer deciphered the dance language of Asian bee species.

Niko immediately tries to talk to the Professor. He has a long list of questions that he would like to discuss with him. Especially to Niko's most important question, where he can find giant honey bees in Sri Lanka in winter, Professor Lindauer does not know the answer: "I have found many *Apis dorsata* colonies in the highlands of Sri Lanka. The people there have assured me several times that the bees leave this area in August and September. In winter, there are no giant honey bees there. It is then too cold in the mountains. Nobody knows where the colonies are migrating."

He gives Niko the address of Dr. Bruce Baptist at the University of Peradeniya, who helped him almost twenty years ago: "Dr. Baptist is the most renowned expert on honey bees in Sri Lanka." It's a very short conversation with no clear recommendations. Lindauer apparently had difficulties on site. He speaks several times of the lack of reliability of his partners. Niko obtains all the travel guides and travel literature about Sri Lanka that he can find, from the old Ernst Haeckel in 1881 to the last publication of the Peradeniya University of Dr. Fahrenhorst on the chromosome number of the *Apis* species. A detailed letter to Dr. Baptist asking for help is sent by airmail to Sri Lanka.

Gudrun is "not amused" about "Niko's hectic actions", as she says. He lies awake for a long time at night and also disturbs Gudrun's sleep. Where are the colonies of *Apis dorsata* from November to March? Dr. Fahrenhorst confirms that they leave the highlands. Where do they fly to? No one can say. A DFG expedition without knowing whether you will find the bee colonies you want to examine and collect?

"Niko, don't do that! It is a mission that cannot succeed. Will you at least listen to my advice now? Apparently, reason and realism are still foreign words to you."

Niko is working feverishly on a new draft for the DFG. He wants to include the study of the migrations of the giant honey bee and requests the support of two students to fetch three colonies of *A. dorsata* for the Institute's new bee flight room. In return, the costs for an off-road vehicle and the long car journey to Pakistan are eliminated. What will a critical reviewer say about a project that stands on such "shaky feet"—as Gudrun and Professor Ruttner express themselves?

Then comes the call from Dr. John: The expert has agreed. Niko can start preparations for the trip to Sri Lanka—initially at his own risk. The written and thus binding approval of the DFG can only be granted in three weeks, after the next committee meeting. "Previous experience shows that such small projects are usually approved without discussion," says Dr. John.

Niko's application to work in Sri Lanka for three months is approved. Gudrun does not see any particular problems for women in Buddhist Sri Lanka. The socialist government under a woman, President Sirimavo Bandaranaike, seems sympathetic to her. After her experiences with the six-week separation because of the Pakistan trip, Gudrun is happy that she can avoid the three-month separation. The plane tickets for Gudrun and the children cannot be paid from the project funds. They have to pay for them from their private savings. "Three months with Anna, Fritz and me in Sri Lanka are hopefully worth it to you?" Gudrun jokes.

"Such a stupid question! A flight even to the ends of the earth would not be too high a price to pay for the joy of being with you. We should take care of the tickets quickly. I suggest that I fly earlier. A longer stay in hotels would probably not be nice for the children. With ten days' notice, I will have found a good place to stay for the children and you. Unfortunately, no outbound flight together," Niko notes regretfully.

"Agreed. The children need a permanent home in Sri Lanka immediately. The transition to the new environment will then be easier for them. A long-haul flight alone with Anna and Fritz is certainly not a walk in the park, but it is always the lesser problem. Niko, you're going to pick us up at the airport, aren't you?" "You can rely on that."

The cheapest flight to Colombo is offered by Aeroflot. With a stopover and a change in Moscow. It starts at the beginning of December (1975). The children sleep on the flight to Moscow, after landing they are lively and enterprising. Anna and Fritz run through the huge halls and corridors and play tag. Gudrun struggles to keep an eye on them and at the same time find her way to the gate for the flight to Sri Lanka. They pass a restaurant, elegantly dressed in white cloths and cloth napkins. Will they be able to go in there? There are still more than two hours left until departure, so Gudrun sits down at one of the tables. Up close, she sees the many stains on the white tablecloths. Are Russian detergents that bad? She promises the children Coke and wants to order a hot tea for herself. No waitress comes. After almost 20 minutes, Gudrun takes the children by the hand and heads towards the kitchen. Several waitresses and waiters sit there and fill out some papers. Gudrun calls and asks for service. An older woman comes, looks at her grimly and says: "Nijet". Does she not understand English? Gudrun doesn't understand Russian, but Nijet is unmistakable. She answers with "Nijet, Nijet, Nijet" and demonstrates in sign language that the children are thirsty. Then a man joins them. Obviously a superior of the woman. His English is incomprehensible, he hears Gudrun talking to the children. "I'm sorry," he says suddenly in perfect German, "at the moment it's shift change. The restaurant will not reopen for an hour. We are closed and can't officially serve anything. With such young children, I want to make an exception." He shouts something in Russian and his people bring two paper cups of lemonade. "A gift. Don't pay!" Gudrun thanks him and wonders how long a shift change at an internationally airport can take.

The flight from Moscow to Colombo takes eight hours. Fritz sits on Gudrun's lap with his three teddy bears. He has to go to the toilet. Anna sleeps peacefully on the seat next to her. Gudrun gets up quietly, places the teddies on her seat despite Fritz's protests and walks with him to the back. Gudrun is lucky, a toilet is free. Everything happens very quickly. When she returns, what a shock, Anna has disappeared. Fritz starts crying and shouts "Anna, Anna!" Gudrun rings the bell for the stewardess. It doesn't take long and a Russian announcement comes over the loudspeakers. Gudrun understands "Anna". It's a good thing that her daughter has an internationally understandable name. Shortly afterwards, a young woman, probably a Ceylonese woman with Anna and another child, arrives: "Excuse me, madam, your daughter came to our seats a few minutes ago and started to play with my daughter. This is Nilanti, my girl! I did not realize that you were looking for her."

Gudrun only understands half of Ceylonese English, but she has grasped the most important thing: "I am Gudrun Koeniger. Together with my children I am on the way to Kandy, to meet my husband."

"Oh, how nice, we live in Kandy. I am Shereene Fernando. Here is my card. Please call me when you are there. My number is on the card. We must allow our children to continue playing together."

The "Return to your seat" and "Fasten seat belt" signals light up.

"Cheerio!" Mrs. Fernando waves and walks forward with Nilanti.

The rapid descent to the airport causes earache. The children cry loudly and cannot be comforted. Gudrun has to swallow several times to adjust her eardrums to the rapid change in air pressure. One passenger grumbles: "Such a damn fighter pilot. The pilot must have completed his training in the glorious Red Army!"

In the arrival hall, everyone jostles, pushes and wants to be the first at the baggage carousel. Fritz cries and wants to get cling to Gudrun's arm. Gudrun carries a lot of hand luggage, including teddy bears, a large trolley and a not quite light backpack. Anna also clings to her hand. A difficult situation. Here comes Mrs. Fernando. In front of her walks a customs officer in a colorful uniform. With harsh orders, he makes his way through the crowd. "Hello, Mrs. Koeniger, this is Mr. Kanshi. He is a nephew of my husband and he wants to help us through the customs!"

Mr. Kanshi asks Gudrun about the baggage tags, which he then gives to a porter.

When Shereene Fernando's last suitcase arrives, it's off to customs.

"Nothing to declare." Mr. Kanshi makes it clear that there is no contradiction here. They are soon in the arrivals hall. Gudrun looks around searchingly. Where is Niko? He is not to be discovered, but then she sees Jochen, Niko's student, who towers over the crowd at 1.85 meters. Jochen has also discovered her: "Welcome to Sri Lanka! How was your flight?" "Where is Niko? He promised to pick us up. Is he sick?"

Gudrun can't hide her disappointment. "Niko is on the road in the east of the island. He hasn't found any *Apis dorsata* yet and is quite desperate. He promised to be at our house in Kandy tonight. Sorry, you have to be patient for that long. Where's your luggage?" Shereene Fernando and Mr. Kanshi met their driver. Before Gudrun can thank her for the help, Mrs. Fernando waves: "Cheerio, and please do not forget to call and come to our house".

Jochen has come in a taxi, an old Morris Minor. The driver takes over the luggage from the porter, who comes to Jochen and wants his wages. 50 Rupees. Jochen only wants to give 30. The porter protests. Gudrun intervenes: "Jochen, we are dog-tired and want to drive off. If you don't pay full pay right away, I'll give that carrier $10. I don't have rupees." The threat makes an impression. Jochen is offended: "Yes, boss." In the meantime, the taxi driver, who introduces himself as Mr. Chandaratne, has performed a miracle: the four suitcases are lashed to the roof rack of the small car. Gudrun gets in the back with the children. Jochen folds his legs and sits down next to the driver. "We're ready to go."

The beginning of the journey, through Negombo along the coast, is flat. Despite the heavy traffic, the Morris Minor still reaches a speed of about 50 kilometers per hour. Then the taxi turns and drives towards the mountains. Fortunately, the children fall asleep after some whining. The progress is even slower here. The road is narrow and winding. On the climbs, it only continues in second gear. People are walking on the street. There are many people at watering holes, washing laundry, themselves or both. For the Sri Lanka shower, water is scooped in clay pots or plastic buckets and poured over the head. The clothes, the wide saris of the women and the sarongs of the men, consist of long lengths of fabric that are elegantly wrapped around the bodies. Gudrun is satisfied that the driver drives slowly and carefully. She hopes that the children will sleep for a long time. After almost four hours, they turn into Anniwatta Road, where the rented house is located. It goes steeply uphill. Suddenly, the Moris Minor stops. The engine seems too weak to make the last part, the driveway to the house. Jochen gets out and walks. Then the car crawls in first gear. When it stops, they are standing in front of a two-storied stately house with a large garden. Gudrun has to take a deep breath—such a luxury! Niko comes running and opens the car door. Gudrun points to the sleeping children and whispers: "Better not wake them up!" Niko carefully takes Anna in his arms: "You take Fritz." Mr. Chandaratne opens the front door softly. On the right go up the stairs to the first floor. In the children's room, Niko lays Anna on her bed and goes to meet Gudrun. He takes Fritz and carefully lays him on the second bed. Then he whispers "mosquitoes" and attaches the net to Fritz's bed. Gudrun anchors Anna's mosquito net. They sneak out of the nursery, leave the door ajar and hug each other.

"Finally! Come to our room!" Niko pulls Gudrun into the large room next door. Gudrun falls exhausted on the bed, only to get up again: "Where are the toilet and shower?" Niko opens a lattice door next to the wardrobe: "Here's an attached bathroom!" They hear a soft knock and Jochen and Mr. Chandaratne are standing in front of the door with two suitcases each: "We'll bring the bag and the other things in a moment!" Niko thanks him and hands Gudrun a fresh towel for the shower.

"Why don't you take my kimono until we have unpacked your suitcases. It hangs over the bar at the back." Under the cold shower, Gudrun's tiredness evaporates. Niko's kimono is too big for her, but at these temperatures the airy cotton cloth is very comfortable. "Niko, why didn't you pick us up from the airport as promised?"

Thirty-third chapter
Niko searches for the giant honey bees and Gudrun settles in Kandy

"Excuse me Gudrun, but sit down and try the tea. It's excellent!"

Niko pushes Gudrun's cup in front of her.

"Jochen was in Colombo anyway and negotiated with Swissair about transporting the bee colonies to Frankfurt. It was obvious that he would pick you up and I would continue my search for the bee colonies. I haven't found one. I'm desperate."

"Desperate? That doesn't sound like you at all. Are you sick? You must have lost a kilo or two." "That's because of the Ceylonese cuisine. I mostly eat in small rest houses in remote villages. There is only local cuisine. Always rice and curry. Very spicy! I can only get a little down. Still better than the cream crackers with Coke during the first few days. I was on the road a lot."

"You're looking for *Apis dorsata* colonies. How do you do that?"

"Sri Lanka is an island. All the colonies that were in the highlands in the summer must be somewhere else now. I want to go around the highlands and am currently driving along the east coast. I usually stop about every 20 or 30 kilometers. My driver, Mr. Ossen, then asks the people who come to our car curiously if they have seen a bee colony. Unfortunately, and this is my problem, people say "yes" too often. It is probably impolite to say "no" in this country. We then ask for details. Especially when the colonies were seen and how far away the place is. If the information sounds probable and the way is no more than two miles, we set off. Often through dense bushy forests. In the heat, it's hard for me to keep up with the pace of the people. After an hour's walk at the latest, we want to see the bees. Often the informant then points to some tree and says that he saw the bees in it. Most of the time I am sure that there was never a bee colony there. Branches that are too thin or no horizontal ones, which these bees need."

Niko groans: "The first time I expressed my frustration and cursed. All the people were against me. Mr. Ossen has just managed to defuse the situation. We offered our Golden Leaf cigarettes and smoked together."

"That doesn't sound good."

"I've learned my lesson. When, after a long march, exhausted and sweaty, I once again realize that there could never have been bees at the specified location, I laugh out loud and say that the informant probably thinks I'm "stupid". I'll explain why I know that. Then everyone laughs at the fraudster. Peace, joy, pancakes! We smoke together and go back to the car."

"Niko, I need a round of sleep now. My bed is on the left like at home?"

"As you will, Gudrun! Don't forget to close the mosquito net carefully. I'll take a careful look after the children."

Niko goes very quietly into the children's room.

When Gudrun wakes up again, she can't believe her eyes: a cloud floats in through the open windows, moves over the bed and flies out again at the opposite window. And the next one comes. It gets much cooler in the room. How strange.

For the next two days, Niko stays with Gudrun and the children in Kandy. There are no problems with the children. On the contrary, they have a lot of new things to discover. Like their mother, they enjoy the fact that instead of snowsuits with boots, they only need panties and T-shirts–and small straw hats! They quickly discover that many mimosas grow on the long garden path down to the street. As soon as they touch the leaves, they fold. When they reach the bottom, the leaves at the top of the path are unfolded again. They quickly run back up and the game starts again! Or they feed the chickens. A hen has chicks and leads them through the garden. As soon as danger is in sight, the chicks disappear into their mother's plumage.

Unpacking the suitcases and furnishing the rooms is done quickly. The children are busy and not many things fit into three suitcases. They go shopping at the market, where they have everything. The students Jochen and Ralph, who had only stayed in the house until now, have hired the taxi driver Mr. Chandaratne to come by every morning around ten and drive them to the market. This time only Gudrun, Niko and the children are driving. Splendid! What a hustle and bustle! The market hall extends over two floors. It is divided into fish and meat market, incredible quantities of vegetables and fruits. It smells of a wide variety of spices. On the upper floor there are clothes, especially saris and sarongs made of beautiful batik fabrics, but also western trousers, shirts and dresses. Gudrun is fascinated.

Unfortunately, the small blond and blue-eyed children quickly become an attraction. Many want to know what foreign skin and blonde hair feel like. The adults pinch the children's cheeks in a "friendly" way. Soon Anna and Fritz start crying. Nevertheless, it is difficult and apparently also impolite to prevent the touching. That's how it goes when you look different! Gudrun remembers a colleague from Madagascar with pitch-black skin. He had visited them at home in Schwalbach and Fritz and Anna climbed onto his lap. Tears came to his eyes: "The first time that children are not afraid of me!"

In Kandy, it's the other way around. The people, whether young or old, behave so curiously and intrusively that the children cling to their parents, cry and just want to go home. You have to go back to the taxi, even though only half of the groceries have been bought. Mr. Chandaratne says: "I should have been better keeping up and protected the children. And they paid far too much. Give me the list, I'll buy the rest quickly. You stay here in the taxi!"

He has done everything in ten minutes and spent much less money. In the future, Gudrun will only take the children to the market if a local is present. Most of the time, Mr. Chandaratne will only get a list and money and buy everything on his own. This is especially practical for Gudrun, because Niko will continue to be on the road a lot.

The students are busy with experiments on the heat balance of *A. cerana*. Professor Baptist, whose address Niko had received in Frankfurt, placed two of his colonies at their disposal.

Niko and Gudrun had agreed that they did not want to live with servants in Sri Lanka like rich locals. In Germany, they did everything themselves. The household is more difficult to run here. Two students, Gudrun and Niko–four adults and two children–need to eat. Everyone helps, but the rice is mixed with stones and has to be washed in a special fluted bowl. They have to use a device like gold diggers to separate the nuggets from the sand. Then the small stones, which are heavier than grains of rice, remain in the shell. This does not always succeed completely. One day Niko bites a stone and breaks a corner out of a tooth.

Often the power fails and cooking has to be done on an open fire. Gudrun remembers how to do this from her grandmother. Despite all caution, her hands and arms are smeared black, sometimes she even has traces of soot on her face and everyone laughs at her. Through the mediation of the landlords, a domestic helper finally comes. Her name is Sriani and she comes from a family where everyone is unemployed. Her salary is incredibly low from Gudrun's point of view. From her family's point of view, however, Sriani earns a fortune. Soon they are joined by a gardener by the hour, who climbs the coconut palms in the garden with his bare hands and feet and harvests nuts. He keeps the lawn in good shape and is also otherwise available to Gudrun and Niko with advice and support. Mr. Chandaratne, taxi driver and shopper, is the third in the group. Nothing with socialist equality. Gudrun and Niko are rich in Sri Lanka, the helpers earn only a few marks a month. All of them speak English, so they have attended a good school.

At one point, Niko and Gudrun are invited to dinner at Professor Baptist's. The children stay at home with Jochen and Ralph. The food is prepared by Vivienne, Baptist's niece, and tastes excellent, the chili spiciness is just bearable. Niko politely thanks him for the good Ceylonese food. Professor Baptist's answer is embarrassing: "It was supposed to be a European meal." Everyone has to laugh! In the course of the conversation, Niko reports on his still futile search for the giant honey bees. Professor Baptist suggests going to the famous Sigiriya Rock: "It's not on the coast, but in the northern inland. I recently read in the newspaper that tourists were attacked by giant honey bees there. Many want to visit the magnificent palace with rock paintings, which stands on a large rock. The illegitimate king's son Kassapa had it built in the 5th century after he had killed his father in order to become king himself. Fearing his half-brother, the rightful heir to the throne, he has retreated there with his army of thousands of soldiers. Many today believe that these giant bees are the incarnation of the soldiers, because they are there every year. Go there and see if the bees are still there." "Absolutely. I'm leaving tomorrow", comments Niko.

Thirty-fourth chapter

Niko finds giant honey bees and Gudrun is visited by the police

A week after the visit to Baptist, Niko comes home happily beaming: he has finally found the giant honey bees. They do not spend the winter on the coast, but mainly in the middle of the country.

"Dr. Baptist was right! You have to see Sigiriya Rock! In the north, the land is flat. Suddenly, you see a huge boulder towering over the jungle. I can't describe this complex with frescoes, baths, theater and the Lion's Gate to you, you have to see it for yourself! And very close to the famous Lion's Gate, many *dorsata* colonies hang from the rocky outcrops. Unfortunately, you can't conduct any experiments there. They are unattainable. In addition, the palace is a museum and a memorial with many visitors." (Fig. 29, Fig.30, Fig. 31).

"What a shame, that sounds like a wonderful place to work," says Gudrun.

"The best is yet to come! Only about 80 kilometers further north is the old royal city of

Anuradhapura. And many colonies have migrated there as well."

Gudrun is relieved. After almost six weeks of searching, finally the first important result: The bees migrate in winter from the central, southern mountains, which rise up to 2000 meters, to the semi-arid northern central province with an impenetrable bush landscape mixed with small trees. Now the rest of the program can still succeed. Jochen and Ralph are also satisfied. This is celebrated in the evening with a good meal and locally distilled arrack. There is intensive discussion: Jochen, who had set up an flight space in Oberursel before their departure, is worried: "How can we pack the giant honey bees properly so that I can transport them safely to Oberursel?"

Gudrun has to laugh: "First you have to catch them! They don't live in a wooden box on the ground, they hang high in the tree with a large comb on thick branches."

"No", Niko says, "first of all, we have to find colonies that don't hang too high. Then we need a carpenter to ship the colonies in made-to-measure solid boxes."

"How are we supposed to get them in there without the bees stinging us and the comb breaking?"

Discussion goes back and forth. Finally, Niko summarizes the results:

"The colony must hang in the box at the same angle as on the tree. We have to measure the shape of the branch exactly. Then the carpenter has to build a sturdy box with incisions in which the branch fits exactly. The sides of the box must be made of solid material. We hang this box under the colony in the dark and tie it with wire. Then we have to close the box securely on the sides and top with a sturdy material. Only then can we start sawing off the branch. The whole structure must finally be installed in a solid second box for transport by plane and off to Oberursel."

"We have to store the colonies temporarily. Only when we have captured three colonies can we book a flight."

Gudrun: "We need large flight cages for the bees. We can put them up here in the garden in Kandy."

"That can work, I'll send a telegram to Oberursel right away that we need net covering for three cages as soon as possible. And pollen to feed the bees in the cage."

Gudrun suspects: She will now have to stay all alone with the children in the big house for many days. Finding suitable colonies, strong enough and easy to reach, precise measurement, precise construction of the trapping boxes and finally sawing off the thick branches. There is a lot of exploration to be done in the jungle. Not without danger! Not to mention the nocturnal adventure with the sawing off of the branch including the giant honey bee colony!

The carpenter had the frames for three large flight cages ready in a few days and the net covering arrives by express flight directly from Frankfurt. Jochen takes the pick-up to the airport to continue negotiations at customs without the usual help of Mr. Chandaratne, the "house taxi driver". Niko is in Anuradh looking for suitable colonies. When the nets are stretched over the cages and well secured, the two students leave. In four days everyone is back, with a bee colony in a large box, which is initially parked in the shade. While Niko happily hugs the children and Gudrun, the two students quickly disappear into their rooms:

"We finally have to sleep in!" It is three o'clock in the afternoon. "What's wrong with them?" asks Gudrun in amazement. "Oh, let's have a cup of tea first, then I'll tell you."

Niko also seems tired and strained: "You can't imagine how sensitive they are. It started with the fact that the beds in the hotel were not good enough for them, that the mosquito nets had holes and above all that it was too hot for them. I spent all my time in such hotels on my search trips, sometimes in much worse ones." They spent almost the whole night in the jungle to get the bees.

At dark Gudrun wakes up Niko as planned. Together, they remove the packaging material from the bee colony. Everything is placed on a horizontal platform. Then they check whether the protective suits are completely tight. All the lights in the house are extinguished, then they carefully peel off the fabric from the top and sides. This is easier than expected. Half of the bees have clung to the fabric; the other half is still sitting on the comb. In the dark, only a few can take off. Gudrun and Niko carefully place the fabric in front of the comb so that the bees can climb onto the comb from there. Less than five minutes later, the first bees have found their way and are running up. The children stand outside in front of the cage and watch intently. After carefully checking that no bee is still sitting anywhere on the protective suit, Gudrun and Niko leave the cage.

"Let's go to bed and see what it looks like tomorrow morning."

When the students reappear the next morning, Ralph explains: "You must not go to Anuradh with the children because of the beds and the broken mosquito nets. The water from the pipes stinks. It probably comes directly from the huge water reservoirs and massive lakes. There, the water has risen again due to the monsoon rains and as a result, trees, grass and leaves are rotting. In these tanks, people wash clothes and also themselves."

Jochen adds: "And the heat and the constantly yapping dogs and the bleating cattle. Everyone and everything runs freely through the streets; the cattle eat the newspapers flying around and the dogs eat all the garbage. I didn't want to eat anything."

"The curries are sterile, they are cooked for hours," says Gudrun.

"Extremely spicy. I couldn't stand that. I only ate toast, at the beginning with Coke. Coke was soon sold out." "You forget about the masses of mosquitoes, especially in the evening when we had to fetch the bees. They are so small; they fly into the bee veils. Then you can't even beat them to death anymore."

Only now do Ralph and Jochen notice that the bees in the cage have been unpacked and have clustered on the comb. "When did you do that? We thought that we would only settle them today!" Niko answers: "After so much effort, I can't take any risks. Besides, we couldn't have done it in the light! Or did you want to get a lot of stings again?"

The students do not want to go back to Anuradh to catch the next two colonies. Ralph supposedly has to investigate the temperature distribution in the *A. cerana* colonies in the garden, Jochen wants to take care of the captive colony. Gudrun is sure: From now on, she will no longer have to listen to complaints Niko experiences from his great adventures in the jungle every day, while they have to stay in Kandy and have to carry out boring experiments.

Niko finds enough local helpers in Anuradhapura. After ten days, they have the three colonies they want to send to Germany. The servants and visitors are afraid of the defensive giant bees. Almost all of them have been stung at some point and tell how they ran as children while the bees chased them for several kilometers. Even if they had

Fig. 29. The Sigiriya Rock a famous historic tourist sight and home for many *Apis dorsata* colonies in winter.

Fig. 30. *Apis dorsata* colonies nesting on the outcrops of the Sigiriya Rock.

Fig. 31. A cage near the Lion Gate as shelter for visitors in case of defending *Apis dorsata* guard bees. Dramatic happenings! When hundreds of stinging bees are buzzing around, people inside try to shut the door while the people outside would try desperately to open and enter the cage.

hidden in the bush or behind the house, the bees would always find them. Gudrun and Niko decide that there must be a warning about the bees. They get a nice board and write in English and Sinhalese in large letters: Caution! Danger! Colonies of the giant honey bee live here. The board is also painted with bees and a fleeing monkey. The children and domestic staff are happy. The sign is placed at the bottom of the entrance to the garden (Fig. 32, Fig. 33).

One day, a large Peugeot 505 car pulls up in front of their house. Gudrun hurries down the steep path. Shereene Fernando sits in the car: "Hello, you haven't contacted us yet. Asking many friends, I finally found out where you live," she says. "You don't have a telephone, so I came. I want to invite your whole family to dinner on Sunday night." Gudrun reacts ashamed, she had taken the invitation at the airport as a courtesy, not as a serious invitation.

The Fernandos' house is a large and beautiful villa. It is located in the middle of the city center, near the famous Temple of the Tooth, where the tooth of Buddha is venerated as a relic. They are warmly welcomed in the entrance hall of the house. Shereene wears a beautiful silk sari, her husband a colorful silk shirt and Western pants. The Fernando family is a long-established and highly respected in Kandy. Like Professor Baptist's family, they belong to the "Burghers", a group that arose from mixed marriages of European settlers of the colonial era and were entrusted with special tasks by the colonial masters. "Please come in!"

The rooms are spacious and furnished with carved antique furniture. Even before they sit down, a nanny comes with Nilanti and two other children to pick up Anna and Fritz to play. They run happily enthusiastic into the garden. Then a servant arrives with exquisite tea cups and a delicious tea. Shereen wants to know everything: "How are you? Have you found good servants? I would have liked to help you settle in." "Thank you very much, but after a few initial difficulties, we have now managed everything very well. Unfortunately, we won't stay here that long, in six weeks we have to go back to Germany–the University is calling." "Oh, you also work at the university? That's a lot of work. Do you also have a good nanny in Germany? I'm so lucky with my nanny. She is nice and reliable. I can go to work without any worries."

Gudrun thinks to herself: This is the explanation why there are more women at University in India and Sri Lanka than in Germany. German women would have no problems to work if they had nannies, servants and cooks. She says aloud: "Full-time nannies are expensive in Germany, they would earn almost as much money as Niko as a lecturer. We have found a private kindergarten, so I have been able to continue my research for half a year. Almost three years of being just a housewife and mother was unexpectedly difficult for me." Shereene nods: "I can well imagine that, I would certainly have felt the same way. During my trip to Europe, I also noticed that assistants are expensive. Now let's eat!"

On the veranda there is a table with fresh fruit juices, cola and water, as well as several cocktails with arrack, the rum made from the coconut sap. Despite her drinking tolerance, Gudrun prefers to be careful with cocktails. You never know how much alcohol is hidden behind the fruit taste and sweetness.

There is rice with various curries of okra, jackfruit and other vegetables, shrimp and fish with stuffed and crab with a crispy crust. The table is almost bending with delicacies. The children have difficulties with the spicy dishes. Gudrun and Niko are ashamed that they have brought nothing with them except a few old Lego bricks for the children– from which Anna and Fritz have found it very difficult to part with. Shereene must have bought brand new Legos on her tour of Europe six weeks ago. An impressive evening. So much wealth and such luxury does not correspond to their ideas. "How does that fit in with a socialist government?" thinks Gudrun.

The farewell is very warm. "Please, come back soon. And good luck with the giant honey bees! Just be careful, the other day the newspaper said that these bees stung a buffalo to death," Shereene tells them.

The students Jochen and Ralph are in Colombo to prepare the transport of the bees, Niko in Anuradh. Gudrun had told them before that their visas had expired and that they urgently needed to apply for an extension:

"You have to be there and sign in person."

"Oh, there's still time until next week. Then I can stay here longer," was his answer. Gudrun is worried, but she can't do anything alone with the children.

In the late morning, she hears several male voices and looks out of the window. Three uniformed men walk up the steep path to the house. At the front of the gate to the garden they stop and shout very loudly. "Hello, Hello!" Gudrun goes to the gate. She fears that this unannounced visit could have something to do with her expired visas, because these gentlemen wear uniforms. Gudrun still does not fully understand the visitors' introduction because of the typical Sri Lankan English, which swallows many syllables. There a "sergeant" with two subordinates stood in front of her. Gudrun introduces herself and invites the men inside. Her welcome is probably quite timid. The sergeant, Gudrun has not understood his long Sinhalese name either, begins the conversation with the usual question of how she likes Sri Lanka? Gudrun's "Very good" is probably not underlined with the necessary smile. The faces of the visitors become serious. "What about those dangerous bees? We saw your warning board. May we see your bees?" For Gudrun, this interest of her visitors comes as a surprise. The bee issue is more pleasant to her than the expired visas. "Please follow me!" answers Gudrun and leads the way to the flight cages where the *A. dorsata* colonies are located.

The three men look closely and seem surprised. They have a lively conversation in Singhala. Gudrun doesn't understand anything. Then she takes the initiative: "You are interested in honey bees? These are giant honey bee colonies. My husband and his students brought those from Anuradhapura!"

"Why did you catch those bees?" asks the sergeant.

"We want to send them to Germany for research purposes. As you know, these bees can ambush humans and animals. Apparently, they mark their enemies and then pursue them over long distances. This is different from the other honey bee species. All you have to do is wash your hands quickly and you're no longer recognizable as an enemy," Gudrun explains.

"We know when so many bees sting that animals and even humans can die. Why do you want to bring them to Germany?" is the next question.

Gudrun is still glad that the uniformed men don't ask for the passports. That is why she explains everything in detail:

"Our bees also have an alarm pheromone, i.e. a chemical substance that shows the hive mates—this is the enemy you have to sting. We know what kind of substance it is, but it evaporates quickly. The *A. dorsata* use a substance that also serves as an alarm and at the same time lasts much longer. My husband and his students want to isolate this substance and analyze its chemical structure. They can't do that here because of the lack of equipment. Hence the transport to Germany to a chemical laboratory. My husband has an official permit for this."

The men nod, they have understood that. Then they turn to the children, who enthusiastically show them an Agama reptile in a small cage and the chickens behind the house. Then they ask how the bees were captured. Gudrun gives them the reports of her husband and his students:

"The search for the colonies took place during the day, together with the residents who had seen the colonies. However, they can only be packed into the boxes at night. Not even the moon is allowed to shine. Then the bees cannot see anything and cannot defend themselves. It is very difficult and exhausting."

Gudrun is now irritated that the uniformed men still don't ask for the passports. She goes on to say that Jochen has started the first experiments in the cages:

"He gives them either fresh stingers from the small bees or their own and checks whether the bees behave differently. You can see that very clearly. Only, he is also attacked during the experiments. He had a thick net sewn into his protective suit!"

The men have to laugh: "It's a shame we didn't meet him. That was very interesting. Thank you very much." With that, they say goodbye. Gudrun is relieved.

Later they learn from Professor Baptist in Peradeniya that these visitors came from the Sri Lankan secret service. The Socialist Government distrusts the USA and Western countries! Niko's nocturnal activities in the jungle of Anuradhapura had been reported. It was assumed that this was really about mineral resources, oil deposits or other economically important resources that were to be spied on by the FRG on behalf of the CIA. Niko's passport with stamps from Israel, Pakistan and many European countries had further fueled the mistrust of the authorities. Professor Baptist could not convince the agents either.

"One could not and did not want to imagine that you, these visitors from West Germany, are really only interested in bees!" he comments mockingly.

When Niko is back in Kandy, Gudrun insists that they extend their visas by going to the Immigration Service Center in Colombo. The children also have to come along. Mr. Chandaratne drives them as always in his taxi, which takes almost four hours. At least 50 people are standing in a queue in front of the door of the service center. They have to queue up at the back. Chandaratne drives on and looks for a parking space. The heat of the street canyon is unbearable. Anna and Fritz start whining. Then Mr. Chandaratne returns. He explains to Niko that they would have to hire an "assistant" if they wanted to extend the visas today. Gudrun and Niko look at each other, but with the children they have no choice. Mr. Chandaratne sets off and returns after a few minutes with a Sinhalese, the so-called "assistant". He has to be paid, and he needs money to pay the right officials. They hand over all the papers and money to their assistant. Then they wait in a small restaurant and drink well-chilled Coca-Cola. This makes the children happy—otherwise they will never get Cokes! After about two hours, Mr. Chandaratne and the assistant come to them: "You can come with me now." We pass the queue of queues, directly to the counter. Another half hour and everything is done. Relieved, but at the same time with a guilty conscience, they drive back to the cool mountains.

It takes more than two weeks to finalize the transport of the *dorsata* colonies to Frankfurt. This time allows Niko an excursion with Gudrun and the children to Anuradhapura with a long break at the spectacular Sigiriya Rock. Finally, they fly back to Frankfurt. They all—including Anna and Fritz—wish to return to Sri Lanka as soon as possible.

Thirty-fifth chapter

Niko works on Asian giant honey bees in Oberursel and plans another trip to Sri Lanka

Niko spends a lot of time caring for and observing the colonies of the giant honey bee. It is important to optimize the flight room conditions. Day length and temperature are set to Sri Lankan conditions. It takes some time to install the necessary equipment to prevent the relative humidity from dropping below 50 percent. Nevertheless, the three colonies are gradually ceasing their brood rearing activities. Niko is alarmed. Despite an optimal supply of fresh pollen, no more bees are raised. Then the colonies leave their comb and hang as a cluster from the top of the gauze of the cage.

"Maybe this corresponds to the natural life cycle of *Apis dorsata*?" suspects Gudrun.

"A good idea! I haven't thought of that! How is that supposed to work? In the flight room we have almost constant conditions. No natural timers."

"If there are no timers, the so-called internal clock comes into play. Our bee colonies certainly have such an internal clock!"

Fig. 32. *Apis dorsata* colony with high foraging activity inside a flight cage in Kandy.

Fig.33. *Apis dorsata* foragers of caged colonies collect pollen.

"Gudrun, the internal clock leads to a circadian rhythm. This has also been proven in honey bees and results in a rhythm of about 24 hours. Now about three months have passed since we are back in Germany."

"Exactly! I didn't notice anything else. These bees would also start their migration in Sri Lanka at about this time. Is this an annual cycle perhaps innate?"

"Well, maybe you're right in your speculation. What do you think about it: We hang the comb to the bee cluster and see whether they raise brood again?"

"Yes, you should!"

In fact, the colonies are now resuming their brood rearing activity.

"Look, Niko: The colonies have an innate quarterly rhythm!" comments Gudrun.

The experiments with the alarm pheromone of Jochen's giant honey bee work despite the breeding break and are successful. These bees have two active substances, they find out. The first is isopentyl acetate, which has been known to Western honey bees for a long time. It is volatile and immediately alerts the guard bees. The other pheromone, on the other hand, is analyzed for the first time. It is much less volatile and marks the stung enemy for more than 24 hours. This component was probably also decisive for the experience of Professor Lindauer, who twenty years earlier had been followed by the giant honey bees for several kilometers.

Teaching and its preparation also take a lot of time. In cooperation with his colleague and friend Professor Ulli Maschwitz, Niko develops a new concept for the large-scale experimental internship, which takes place at the Bee Institute in Oberursel and not at the Zoology Department in Frankfurt as before. The cooperation leads to a lively scientific exchange with the ant researcher Ulli. At the same time Gudrun and Niko are discussing a further research trip to Sri Lanka. The plan is to carry out field trials in Sri Lanka next winter. They want to settle in Anuradhapura with the children and students. As central topics of the new DFG proposal, Niko focuses on various questions that arise from the common habitat of three honey bee species. Is there food competition between the different sized bee species? Or do the bee species collect different flowers depending on their size? How do the drones avoid cross-mating and find the right, species-specific queen? How do the bee swarms communicate during their migrations? He encloses the official invitation from the University of Peradeniya to work with Professor Baptist with the application.

The International Bee Research Association (IBRA) has invited Niko to a conference on Asian honey bees in London. Gudrun goes along, the children enjoy visiting their grandparents in Wilhelmshaven. In London, Niko gives the introductory lecture on the biology of the giant honey bee. To his great delight, he meets Professor Baptist from Peradeniya: "What a nice surprise to see you here! Can we meet for lunch? We plan to return to Sri Lanka soon and would be happy and grateful if we could then continue the successful cooperation with you."

Professor Baptist answers hesitantly: "I am also happy to meet your wife and you. My trip to London was approved at the last minute after an initial rejection. My students and I will be happy to work with you again. Unfortunately, I can't come to lunch. Can we meet here in the cafeteria later, at 2 p.m.?" "Good! Agreed. See you later!"

At lunch, Niko meets Dr. Eva Crane, the director of IBRA:

"Dr. Crane, I hope my presentation met your expectations?" Niko begins the conversation.

"Fishing for compliments? You noticed from the reaction and the applause at the end that your presentation was very well received!"

Dr. Crane is known and sometimes feared worldwide for her critical comments. Niko is happy about this answer.

"May I thank you for inviting Professor Baptist from Peradeniya? This is a great help for Gudrun and me. We want to continue our work in Sri Lanka soon and can discuss our cooperation with him here. This is much better than correspondence by letter and makes it easier to make arrangements about our experimental work!"

"Professor Baptist had great difficulty getting permission to travel. Sri Lanka has strict foreign exchange restrictions. IBRA financed his flight. I also arranged his accommodation in the hostel of Queens College. That's not expensive. I have the impression that he doesn't have enough money."

"We have an appointment with him after lunch," says Gudrun. Is Baptist's cancellation of the joint meal due to a lack of money? She thinks. "We will take care of the problem. Professor Baptist is a good friend and has helped us in Sri Lanka more than once. Niko and I are happy if we can return his favors!"

"I'm glad to hear that!" replies Dr. Crane. "Baptist is such a wonderful old-fashioned gentleman. He embodies more British tradition than many people here in England, the motherland of the former colony of Ceylon!"

In the cafeteria, Gudrun asks the Sri Lankan professor directly about his financial situation. This violates Ceylonese politeness and causes great embarrassment. He is annoyed! Niko finds a better start and inquiries about the foreign exchange regulation in Sri Lanka. Professor Baptist, no friend of the socialist government of Mrs. Bandaranaike, explains that it is almost impossible to exchange dollars or English pounds.

"An ordinary mortal has no opportunity to legally get foreign currency in Sri Lanka. Our rupees are not accepted outside the country! There is no free exchange. Only party officials and the super-rich have bank accounts abroad. All others are dependent on invitations from relatives or friends who work abroad. A really shameful situation!"

"In your case, we can easily circumvent these difficulties. Thanks to your letter of invitation, we will come back to Sri Lanka next winter and need a lot of rupees there. Please, Dr. Baptist, tell us how many pounds you need. We want to give you the money very gladly!"

"How can I pay back the money," he objects timidly.

"No problem! In Sri Lanka, you can give us back the English pounds as rupees, converted according to the official rate for tourists. We only offer an honest loan. How much do you want?"

Baptist looks from Gudrun to Niko. Both nod with a smile and so Baptist grumbles to himself: "50 or 100 pounds?"

Now they finally drink tea in a relaxed manner and talk about the lectures and the research plans in Sri Lanka. Later, Gudrun and Niko exchange 300 Deutschmarks for English pounds. At the next lecture, they sit down with Professor Baptist and hand him a sealed envelope, which he puts unopened in his briefcase. No comment and no reaction. He seems to need the money urgently and has to take it, but he probably finds his situation humiliating.

Back in Oberursel, Niko is eagerly awaiting the DFG's response. An informal telephone inquiry by the restless Niko is by Dr. John as follows: "Please be a little more patient. You have to wait for the next expert meeting in a week. Then you will be informed in writing. I understand your plan. I would also like to spend the coming winter in Sri Lanka. So much in advance, both experts' reports are extremely positive."

Ten days later, the written approval arrives. The travel preparations are running smoothly. In October 1977, Gudrun and Niko set off with Anna and Fritz. Together, the long journey over Moscow is much easier than the previous flight.

Thirty-sixth chapter

Gudrun and Niko arrive in Anuradhapura and rent a house with a watchman

At the airport in Negombo, the driver Mr. Ossen is waiting with the old Peugeot 404, which they know from their previous stay. "Welcome Sir, and welcome Madam! Where do we go?"

Niko laughs and asks Mr. Ossen about his family. Whether his wife agrees that he can go hunting for giant honey bees again?

"No problem! I have been home for a long time. My wife is happy that she can have the house without me", he jokes.

The suitcases are stored on the roof rack, the trolleys end up under the tailgate. Mr. Ossen negotiates with the porters: 60 rupees!

Gudrun and the children make themselves comfortable in the back seats. Niko sits down in front in the passenger seat. Mr. Ossen has closed Niko's door with a loud bang, he climbs into the driver's seat and asks again: "Where do we go?" Gudrun leans forward: "Mr. Ossen, we need water, cola, bananas and cream crackers first. Please stop in Negombo for shopping. Then we want to go to Anuradh. We will say in good time whether we might need tea and lunch in Puttalam." "Yes, Madam! Good, I will stop at the market in town," Mr. Ossen replies and drives off.

It is dark when they arrive in Anuradhapura. Gudrun and the children have slept for the last hour and now, when the car stops, they wake up. "Where are we?", Anna asks, "Are we there yet?"

"Yes, we are in Anuradh. We will stay here at the Nuwarawewa Resthouse for the first few days until we find a house," Niko answers.

Mr. Wejasena, the hotel manager, greets them warmly. A large room has been prepared: "We have provided two additional beds for the children. Mr. Tillakaratna ordered this for you. Do you know how long you will stay with us?"

"Thank you very much! We are happy to be back here." Niko has his broad Sri Lankan smile on his face: "I think we'll definitely stay for two or three days. We are looking for a large house to rent, for five months. Do you want to call Mr. Tillakaratna? If he has time, I would be grateful if he can come over. I would like to meet him today after dinner around 9 p.m."

Niko walks behind Gudrun and the children to their room. It is much too cold in there. Gudrun turns off the air conditioning and starts the large ceiling fan instead.

"Are you hungry or do you want to continue sleeping?", Gudrun asks the children.

Anna is lying on her bed and Fritz has thrown the teddies on his and lies down next to them. Gudrun and Niko put the mosquito nets under the mattress and check that everything is tight.

"I'm hungry and looking forward to rice and curry. Will you come with me?" asks Niko.

"I'd rather sleep. Also, we shouldn't leave the kids alone on the first night here! When you talk to Tilla: Please do not forget to ask him for a housemaid. I certainly need help with the household! I am sure there are still stones in the rice!"

When Niko has finished eating, Mr. Tillakaratna comes in. Niko is surprised at how slender and slim he looks. He had completely forgotten that. Tilla seems a little embarrassed, but with a broad smile on his narrow face:

"Hopefully I didn't disturb your dinner. Should I wait outside?" he asks quietly.

Niko stands up and greets him warmly:

"Thank you very much, Mr. Tillakaratna, for coming at this late hour. We could go sit down at the bar. We can talk better there!"

They are sitting comfortably in rattan chairs in a corner of the bar. On the table are ice cubes, a large bottle of soda and two well-filled glasses of coconut arrack. After the questions about the news of both families have been exchanged, Tilla gets down to business:

"I have found a new house in the neighborhood that probably offers enough space. The director of the library built it. However, only the ground floor is finished yet. The rent is not cheap at 1000 R per month. There is a large garden for the bees. Dr. Niko, you should look at it tomorrow."

"That sounds very good! Tomorrow at noon? Does that suit you and the owner?"

After this good start, Niko and Tilla order a second arrack.

"Dear Tilla, Mr. Ossen and the Peugeot are here again. In addition, this time I need an assistant who is familiar with Anuradh and the surrounding area and is reliable. As you know, we often have to work at night and overall it's a strenuous activity."

Tilla rocks his head back and forth, takes a sip of arrack and says:

"Dr. Niko, I can help you with that too. A former classmate of mine, Mr. Sellaperuma, is currently unemployed. His English is excellent, he has a good demeanor and knows his way around here very well. Mr. Sellaperuma is your man. I will ask him to come here tomorrow at ten o'clock. Is that good?"

"That's very good! I look forward to meeting Mr. Sellaperuma tomorrow! I almost forgot my most important assignment. Dear Tilla, my wife asks if you could recommend a domestic help? We want to hire a maid."

"Sorry, I don't know any girl. Would you like to ask Sita, my wife Sita. Your wife Gudrun may visit her tomorrow. I think Sita will be able to help!"

Suddenly, an angry elderly lady enters the bar. She complains loudly that she can't sleep because of the loud barking of the many dogs around the hotel. Her English reveals that she comes from Germany.

"Excuse me, maybe I can help you?" Niko speaks to her in German. "Ah, you speak German. I arrived here with Neckermann's group, dead tired, and now this. The dogs bark so loudly that no one can sleep. Make sure that this stops immediately. The dogs also need a night's rest!" "Well, that's a lot of dogs. It will take time for everyone to be calm.

I promise you; the dogs will stop barking." Niko remembers that the dogs usually sleep in the early hours of the morning and continues: "You will have to be patient."

The lady calms down a bit: "Well, then I'll try to sleep again. How long will it take for the barking to stop?" "I'm doing what I can, but it's going to take a while."

Niko has trouble staying serious. When the woman is gone, Niko explains the problem to Mr. Tillakaratna. Tilla laughs: "Unbelievable!"

On the first morning, Gudrun and Niko sit together over a second cup of tea. The children bathe in the pool. "Wonderful here in the warmth! When I think of fog, cold and darkness in Oberursel. A great privilege, your research on tropical bees. Look how Anna and Fritz play in the water!"

Then a waiter comes and tells Niko that a Mr. Sellaperuma wants to talk to him. Niko gets up and follows the waiter. In the reception is a gaunt man, who is small even by Sri Lankan standards, certainly only one and a half meters tall, perhaps around 40 years old. His narrow face is clean-shaven, two incisors are missing, others are half broken. His light-colored shirt and dark trousers appear patched, but clean and ironed. As Mr. Tillakaratna had said the night before: "A former classmate who hasn't had much luck in his life so far."

"You are Mr. Sellaperuma? Your friend Mr. Tillakaratna recommended you. Do you really want to work with us? This is hard work with stinging bees. At night in the jungle, it will take a lot of effort!"

Referring to Ceylonese conditions, he adds: "No office hours. We have to work whenever it is necessary for the bees."

Mr. Sellaperuma now seems even smaller and more reluctant than before. He regains his composure and answers clearly, almost desperately:

"Yes, sir."

He will work hard, and he wants to earn the wages that Tilla has promised him honestly, "penny for penny!"

Niko asks Mr. Sellaperuma to follow him. They go to the pool, where the children are still frolicking, and sit down with Gudrun. "Tea or Coke?" Mr. Sellaperuma can't decide.

"Coke for Mr. Sellaperuma!" orders Niko.

Then Gudrun and Niko discuss their plans. Gudrun will take care of the furnishings of the house.

Gudrun says: "After viewing the house this afternoon, I have a list of the things we need. Mr. Sellaperuma, I would like to go to town with you and our driver, Mr. Ossen. You will show me the shops where we can get furniture, beds and mattresses. We also need a large refrigerator. Is it possible to buy used things here? We're only here for five months, so not everything has to be new!"

Mr. Sellaperuma nods eagerly and smiles. "Yes, madam, I know all the shops and know where we can buy the things you need at a reasonable price. You can rely on me," he replies eagerly and carefully takes a first sip of his Coke.

Gudrun and Niko pick up Tilla at home for the house viewing. Fritz and Anna stay and play with Tilla's sons Kantschi and Janthe. With Tilla they walk across the street to the new house, where the owner greets them. Only the ground floor is completely finished, with a very large living room, kitchen, bedroom and children's room. In addition to the kitchen, there is a small room for servants and a pantry with a large padlock. The roof and also the staircase to the upper floor have also been completed, the upper rooms are still under construction.

"Ideal for our experiments," Gudrun says quietly to Niko.

The house includes a large overgrown area in the garden. In one corner they discover an old unshaven man and in dirty clothes.

"Who is that?" asks Niko.

The owner answers hesitantly: "This is Mr. Samarakoon. He lives here."

He points to a miserable wooden shack away from the house. "Mr. Samarakoon is guarding the construction site. If you want, I can send him away!"

Gudrun reacts indignantly and says to Niko in German: "We can't let that happen under any circumstances. We, the rich people from Germany, chase away a destitute Ceylonese because we want to rent a big house!"

Niko nods and Gudrun turns to Tilla: "We certainly have a lot of work for a capable guard and someone who takes care of the grounds and tends to the plants. Mr. Tilla, can you ask Mr. Samarakoon if he wants to work for us?"

Gudrun had learned that in Sri Lanka nothing can be regulated directly by them. Tilla, the homeowner and especially Samarakoon express their approval of Gudrun's proposal with their very broad smiles, which go beyond the usual extent.

"Yes, we need a capable caretaker and gardener. If Mr. Samarakoon wants, he can work with us. He will get a room upstairs in the house. We want to use the wooden shack for equipment or bee colonies," emphasizes Niko.

They also quickly come to an agreement for the rental of the house. Tilla has clarified the conditions with the homeowner. Niko signs the lease and pays the first rent. In the evening, Mr. Samarakoon comes to the Nuwarawewa Resthouse. He waits outside, holding an employment contract in English and Singala in his hand. Niko pays the first month's salary negotiated by Tilla. Mr. Samarakoon thanks him very warmly. In addition to many Sinhalese words, Niko hears "Thank you" more than ten times.

For the next two days, Mr. Samarakoon is nowhere to be found. Gudrun and Niko discuss what could have happened? "Did he squander the money?"

Niko argues: "For a person who has suffered poverty and deprivation, the temptation should be great to hit the ground running now." Gudrun disagrees: "Samarakoon knows that Tilla and also the homeowner vouched for him. I'm sure Samarakoon didn't drink the money. Hopefully nothing bad happened to the poor old man!"

On the third morning, Samarakoon comes home. He has dressed himself: A new white sarong, a white shirt and red sandals. He is freshly shaved and bathed. He smiles proudly and doesn't look as old as before, more like a sprightly man around 50. Gudrun turns to Niko with satisfaction. "Do you see this transformation! My assessment of Samarakoon's disappearance was correct. He has invested his money well!"

Mr. Samarakoon takes his job as a janitor seriously, chopping a five-meter-wide strip around the house free and sweeping it thoroughly twice a day. This prevents snakes or rodents from entering the house. More than ten keys are hanging from his belt, although nothing is locked in the house. And after four weeks, he hires a young man on an hourly basis to work for him. Visitors then ask Niko where he found this handsome, reliable guard and caretaker.

In the meantime, Gudrun has spoken to Tilla's wife Sita, who works as a nurse in the hospital, and told her about her bad experiences during her last stay in Sri Lanka: "Ultimately, cooking on an open fire is an art that I have not mastered. In the end, everything was black. Not only the tables, the cupboards and the kitchen, but also myself!"

Sita laughs and knows that the soot from the firewood is very difficult to clean. She knows a family with an unemployed daughter who speaks good English.

"She is a reliable and honest girl, older, about in her mid-20s. Her name is Indrani and she has helped out in our hospital. I'll talk to her mother and send Indrani over to you!"

The next morning, as they are putting away furniture, Indrani stands at the door, a slim, shy young woman. She asks Niko about "Madam". Gudrun is supervising the setting up of the beds. She shouts: "Tell the girl she has to wait now. The mosquito nets and beds are more important." Niko tells her that Madam doesn't have time at the moment. "Please, Indrani, wait five minutes!" Indrani reacts frightened, goes outside and waits. Not long after, Gudrun rushes to her, sweaty, and is greeted by Indrani with a curtsy. Gudrun is in a hurry. She explains to Indrani in short words what housework and duties she will have to do. Indrani confirms this point by point with "Yes, Madam." Then Gudrun asks, "When can you start? Tomorrow morning around seven?" "Yes, Madam!" The next day Indrani does not come. Gudrun talks to Sita at noon, who in turn sends her servant to the house of Indrani's family. Indrani's mother then visits Sita in the hospital and explains to her what had happened.

"Madam never laughed, always just said loudly and sternly what her tasks would be. She is good at cooking and cleaning."

Sita immediately understands the situation and summons Indrani to her. She patiently explains to her that the Germans have different customs.

"They speak loudly and don't laugh. They are still friendly and nice. Once you're there, you'll understand these people better. Believe me, this is a good job and very good pay. You will love the children and the parents! If you want to know more about these people, ask Mr. Samarakoon. He's been there for some time and feels comfortable!"

Finally, Indrani anxiously agrees to try this German family, despite the bad first impression. Gudrun is shocked when she finally learns how much she has frightened Indrani. Without a smile on your face, nothing works in Sri Lanka. For Gudrun, probably because of her North German upbringing, this is a great challenge!

Mr. Sellaperuma proves himself in procuring the furnishings for the new house. Gudrun explains to him in detail what needs to be obtained next. Sellaperuma knows the right shops in Anuradh, Gudrun quickly finds what she is looking for, decides what to buy and leaves the complicated price negotiations to Mr. Sellaperuma: The salesman smiles and Sellaperuma smiles without interruption. It goes back and forth. Usually longer than ten minutes. Gudrun admires the fact that everyone always manages to save face despite the often long negotiation.

Indrani proves to be extremely helpful in the furnishings, and Gudrun tries to praise her often with a smile. Indrani gets the small room next to the kitchen.

After three days, everything is more or less complete. Gudrun, Niko and the children leave the Resthouse and move into their new home. However, not without the promise to go to the swimming pool regularly.

An ancient English car–probably a pre-war model–stops right in front of the house and blocks the driveway. Mr. Ossen presses the horn of the Peugeot. Then Gudrun recognizes Professor Baptist, who gets out stiffly and a little stilted with his scraped briefcase under his arm.

"Don't honk, Mr. Ossen! That's an important visitor."

Gudrun jumps out of the car and walks towards the Professor: "Welcome, Dr. Baptist. Sorry for honking, we're in the middle of moving and didn't recognize you right away. This is our new home here in Anuradh," she apologizes. Professor Baptist greets them perfectly and warmly. Then he mumbles something about "Blocking your entrance" and waves Mr. Ossen over, to whom he hands his car key. He probably asks him on Singala to drive the car away from the driveway.

"Tea or water, what can I offer?" Professor Baptist replies that he doesn't want to make any fuss and has just had a cup of tea at the Resthouse. There he was told where the Germans could now be found. Gudrun goes into the kitchen and asks Indrani for two glasses of cold soda water.

She sits down with Professor Baptist: "Niko is in town and meeting some people there. He wants to find out where he can find colonies of giant honey bees. And I'm the poor housewife who furnishes the house," she laughs.

"I see you two have a lot to do. I don't want to disturb."

Baptist clears his throat, opens the old briefcase, and places a large envelope and paper on the table: "These are the rupees I owe you. Here's the conversion. You gave me 100 English pounds in London and converted to the current exchange rate for tourists, that's 3,075 SLR in the envelope here. Please count, Dr. Gudrun. And by the way, thank you very much. Without your credit, it would have been difficult for me in London. You don't even know from what misery you have saved me."

Indrani has added ice cubes to the soda water and placed a slice of lemon on the plate next to the glass. She bows and disappears with the tray.

"Where did she get the lemon?" Gudrun thinks. She will ask later. Gudrun opens the envelope with the money: "Thank you very much! That must have really taken time. We are here for a few months we will surely be in Peradeniya sooner or later. May we visit you there?"

Gudrun's direct answer surprises Professor Baptist. He takes a first sip of soda water, probably to bridge the resulting break.

"Two of my students, Mr. Wijeyagunesekara and Mr. Punchihewa would like to work with you. When should they come?"

Gudrun suggests the beginning of next week. "By then, the preparations will have been completed and we can start with the projects. By the way, we are also expecting two students from Frankfurt. I'm looking forward to a full house."

Professor Baptist thanks her again and they go to his car.

"Don't you want to wait for Niko? He has to come any moment."

"I will stay with Asoka Mahadiulwewa here in Anuradh. Please come by with Niko in the evening."

He waves and Gudrun hears his croaking voice: "Cheerio!" Gudrun thinks about how she could find out where the Mahadiulwewas live? Then she remembers: Mr. Tillakaratna or Mr. Sellaperuma, probably even both, will know.

As soon as they have moved in, Anna and Fritz start scratching. In the morning after getting up, they have a particularly large number of small swellings, especially on their abdomen. Anna scratches herself a lot and Gudrun has to be careful that Anna not to scratch herself and cause sores. At first, they think of mosquito bites. Gudrun closes the mosquito nets particularly and carefully. All corners are plugged up. "Niko, please check again if the mosquito nets are really tight!"

"Yes, there is definitely no mosquito in there. The children can sleep safely."

Then Gudrun and Niko also get these swellings in places that are always protected from mosquitoes. They talk to Sita, who knows her way around as a nurse. A look at Anna's belly is enough for her. The diagnosis is: "These are bites of bed bugs. Have you taken over the old beds?"

Gudrun shakes herself: "Bed bugs!"

Sita recommends: "The best way to get rid of these bloodsuckers is to put all mattresses and bedding in the sun every morning. Turn over frequently. Then the bugs look for another place."

Fortunately, Anna did not understand the English conversation. Niko and Gudrun decide not to tell the children about the bugs. In the morning, the mattresses, pillows and blankets are spread out in the sun. After three days, the spook seems to be over, only Anna still has sores. Niko takes down the Ceylonese mosquito net, which consists of a bamboo ring at the top, from which a wire extends to attach it to the ceiling. Downwards, the mosquito net hangs down like a four-poster bed. Between the bamboo ring and the net, they find what they are looking for–more than ten lively bed bugs. This time it is treated with a "chemical club": Baygon can be bought in a shop. Then there is silence. However, the daily airing of bedding in the sun will continue.

Professor H. Levinson, a friend and colleague, writes them a letter after he has learned about the bed bugs—probably through a letter from Professor Maschwitz. "I urgently need bed bugs for my experiments and cannot get any here in Frankfurt! And you had some and killed them all instead of sending them to me!"

Mr. Sellaperuma hears Gudrun and Niko talking about the letter and smiles shyly: "You think you killed all the bugs?" He takes a chair from the dining table and knocks it on the stone floor with force. And lo and behold, some bugs tumble out of the cracks. After they have knocked open all the chairs well, they find 21 bed bugs that are packed up and sent to the laboratory in Germany by airmail letter.

Thirty-sixth chapter

Gudrun and Niko celebrate Christmas under a palm tree

In mid-December, Professor Baptist's two students arrive. Noel Wijeyagunesekara is a small and very slim young man. The long hair that frames his narrow face with the prominent nose is unusual for Sri Lanka. Overall, he makes a reserved, even shy impression. Wasanta Punchihewa is also no taller than 1.60 meters, but stocky, has broad shoulders and seems to be quite used to field work. His appearance is very determined. He speaks in a loud voice and seems confident. He is the spokesman for the introduction: "This is Noel Wijeyagunesekara, who has just completed his master's degree in Peradeniya and is now waiting for the final approval for his scholarship to study biology in England. Noel works as a course assistant to Professor Baptist. My name is Wasanta Punchihewa. You can just call me Punchi. I come from a family of plantation managers. My father was a planter on one of the large tea plantations in the highlands. Professor Baptist said I could learn a lot about field research from you here, and so here I am!"

Gudrun takes on the role of hostess and invites the students in. Indrani, who has obviously heard everything, comes over and asks if she should bring tea. Gudrun thanks her and says yes. "A glass of water is enough for me!" Punchi intervenes.

"So only three teacups and water for everyone, please."

After a short tour of the house and the garden, they sit down at the living room table. Niko outlines his research program. He suggests that Noel start with the different honey bee species when determining the flight times of the drones. Punchi, on the other hand, could perhaps help with the work on food competition?

Punchi objects. "I would rather take on a research task that is independent of you. I want to continue when you return to Germany,"

Niko is annoyed, then Gudrun intervenes:

"Punchi, I can understand your point of view. Training on an artificial feeding place—we planned this to observe the competition between the three differently sized bee species—is really difficult and requires the cooperation of other people. We have many other topics. Niko wants to investigate the average flight distances in search of food. You can do that well on your own. You have to observe the dances of the different species and measure the flight distances exactly. Do you want to do that?"

Punchi asks for time to think it over. He then wants to discuss more details about this topic with them. Then Noel speaks up:

"Of course, I would like to help though I can't judge Niko's proposal about drone flight yet. For me, it's important that we finish this project before you leave and before I go to England."

"Yes Noel, I think that we will finish this project quickly," Niko answers. "We should meet here again tomorrow afternoon. If all goes well, Sabine and Michael will also be here. Our students from Frankfurt landed in Colombo yesterday and are coming to Anuradh by train tonight."

Noel and Punchi say goodbye. With Tilla's help, they look for an accommodation in Anuradh.

Gudrun looks at Niko expectantly: "The good (Professor) Baptist sent us two special specimens. Two students couldn't be more different than Noel and Punchi! Have you ever met a German biology student who speaks as softly as Noel? And Punchi annoyed you with his rude reaction."

"Well, at least we got a first impression. Punchi's reaction was quite brazen! Would you like to help with his care? You'd be doing me a big favor with that. In any case, we should make the most of the help of these two students. As children of this country, they have many advantages and opportunities that we lack."

Finally, Gudrun finds a solution for the children's lessons with Sita's help. A retired major of the army takes over mathematics and an unemployed teacher gives English lessons. Gudrun has received corresponding textbooks and documents from both Anna's and Fritz's teacher, which she has supplemented with English remarks. In spite of all the initial difficulties due to the children's insufficient English skills at the beginning, this private lessons are going surprisingly well and is fun for both the children and the teachers. In the afternoons, they play with Tilla's children of the same age. The different languages are only a minor problem. The Tilla boys are particularly enthusiastic about the Lego bricks Anna and Fritz have brought from Germany. The four build imaginative castles or small villages with cows together. Of course, Kanshi and Janthe also want to show their parents what beautiful things they have built. Here, Mr. Samarakoon proves to be a "grabber" janitor. He checks the boys as soon as they want to leave the property. Even the trouser pockets have to be emptied. Anna and Fritz hear that their Sri Lankan friends are crying and call Gudrun for help. "The children are friends and are allowed to borrow the toys. They want to bring everything back tomorrow," she explains to Mr. Samaracoon.

Overall, Mr. Samarakoon takes his job as a guard for the inexperienced Germans seriously. He also catches the servant of the neighboring house when he wants to get in to Indrani through the window at night. With a loud hello, the would-be Casanova is chased away. Tilla's wife Sita later explains the problem to Gudrun and Niko:

"If a maid becomes pregnant during employment, the master of the house must care for the child until it is an adult. Either he didn't pay enough attention or he is the father of the child himself! It doesn't matter how it happened–the landlord pays."

"Now I understand why Indrani asks me for permission to go to cinema at the age of 23. And then our guard always goes with her, I have to give him money for the entrance. So, it is necessary and not just a trick to go to the cinema for free!"

Christmas is approaching and Gudrun is thinking with the children about how she can find a Christmas tree. Most families of the development workers travel to Germany over the Christmas holidays, one family in Kandy orders a cypress from the mountains of Sri Lanka for a lot of money. Gudrun is surprised about this effort (Fig. 34).

"We should have a Christmas tree! The whole neighborhood is looking forward to celebrating a German Christmas with us." "We don't have any decorations for the tree. How is that supposed to work?" asks Anna skeptically. "We have to think carefully about that. First, we find a tree, then we see if there are Christmas decorations to buy here. Otherwise, we have to make decorations ourselves. We will find a good solution. We must not disappoint our neighbors!"

The children grumble and want to play. Then Gudrun has an idea:

"You know what? In Germany, we only have conifers as Christmas trees because they are the only trees that are green in winter. Here there are only green trees. We have to look for some large plant with horizontal branches and green leaves. Horizontal is important so that we can hang the decoration and put on the candles. Come on, let's go to the shopping street and look for a nice big plant in a pot, and also for silver and gold paper."

"Oh yes, are we going right away? And may Janthe and Kantschi come along?"

"Of course, act quickly. I'm treating everyone to an ice cream at the Elefant House!"

At four o'clock in the afternoon, the worst heat is over. It only takes a quarter of an hour to walk to the shops. There is no flower shop in Anuradh, but many owners have decorated their entrance with plants. And suddenly Anna shouts: "There, take a look! Isn't this palm tree beautiful!"

There is an Areca palm tree in a pot, without a firm trunk, the dark green leaves bending horizontally over the pot at a height of a good meter. Ideal for decorating (Fig. 35).

"This is our Christmas tree!" everyone agrees. Fritz wants to buy it immediately, but it's not that easy.

"The palm tree is not for sale here. We have to ask Tilla for help. And besides, we can't carry the heavy flowerpot all the way home."

The three boys want to protest, but the invitation to ice cream lures them away from the palm tree. There are about 20 shops in the street. With ice cream in hand, all the shops are quickly searched for Christmas decorations. Nothing. Anuradh is mainly home to Buddhists and Hindus. There is no glossy silver or gold paper either. Janthe and Kantschi explain in Singhala what they are looking for, that doesn't help either. Red glossy paper is also not to be found. They give up and run back home. The children explain to Tilla that he has to help.

Gudrun is waiting for Niko. Since the beginning of December, he has been on the road with Mr. Ossen in the car again and asking about *A. dorsata* colonies in the villages. It is now almost mid-December and no one has seen these bees. He comes home depressed: "What am I supposed to do? The DFG has approved the money for *A. dorsata* research and I still haven't found one. The experiments with dwarf honey bees may be interesting, but not for the DFG. They want results on the topic I have requested. If I don't make it, who knows if I'll ever get a research fellowship again."

"When did you find the colonies two years ago? Wasn't that just in January?"

"Yes, but everyone told me that they would arrive here at the beginning of December!" "The bees are sure to come, and if it gets later, then we all have to work overtime. We can do that." Niko nods, he is nervous.

Gudrun tells him about the Areca palm: "It's almost as tall as I am and very suitable for decorating. Tilla has to get it for us, it's standing in front of a shop and can't really be sold."

"It's all a question of price. It's nice that you found this palm tree."

"We didn't find any golden paper, nor silver or shiny red paper. Going to Colombo especially for this? You would be on the road for more than twelve hours. And how are you supposed to know if and where you can find it in Colombo?"

"We often smoke Gold Leaf cigarettes. They are particularly popular with everyone. I can buy a few boxes there. I need many to offer individual cigarettes to the people in the village. They are happy and then search more intensively for the *A. dorsata* colonies. I will collect the gold paper with which they are wrapped."

"That's a good idea. I just thought of the aluminum foil we brought for the experiments. I'm sure I'll also find red paper for crafting. Great, then at least the Christmas tree is secured. The first neighbors here have asked about our Christmas tree. They want to see a German Christmas tree and celebrate with us."

"The palm tree will not become typically German," Niko points out.

"Hopefully nice," Gudrun replies.

Tilla manages to get the Areca palm on loan over the holydays. Even without decoration, it is a great enrichment for the large living room. Gudrun finds some colored paper

Fig. 34. Family Tillakaradna and Family Koeniger.

and then they eagerly make decorations. The children have fun and are well occupied with the now frequent rainy days. Gudrun is constantly coming up with new handicraft methods from the post-war period. Balls made of colourful circles that are pinched three times on the sides and can thus be glued together to form balls, colorful chains made of paper rings, angels individually or several holding hands with silhouettes. Indrani and also Janthe and Kantschi help and the neighbors like to come by to marvel at the Christmas workshop. Some also want to do handicrafts. With Indrani, Gudrun goes to the best baker in Anurah and chooses different cakes. They don't necessarily have to be German, they have to taste good to the neighbors.

By December 22, Gudrun has prepared almost everything with Indrani's help. The neighbors are invited for Christmas Eve, and everyone is looking forward to the festivities. Then Niko comes home unexpectedly early at noon and gets out of the car beaming: "Now I have the best Christmas present. The very best. I have received a message from three villages that the first swarms of giant honey bees have arrived. I have to check them immediately and prepare everything to settle the bee colonies in the garden. Christmas is cancelled this year, cancel everything." Gudrun is happy at first, but then she gets angry: "Yes, it's really like a Christmas present for us, but not for Anna and Fritz! The neighbors won't understand that either. For them, holidays are sacred, for January we are invited to the particularly important Duruthu Full Moon Poya festival, the day of Buddha's first visit to Sri Lanka about 2500 years ago—"on the day of the first full moon of the year." I am not canceling our Christmas party under any circumstances. If necessary, we will celebrate without you, Niko. I won't help you either. Please drive to the villages and see if the swarms are really there. Catching them right away and settling them here in the cages, that's not possible."

"The swarms don't stay for long, they move on or they look for a tree or water tower as a nesting place, and no one knows where. What should I say to the DFG then? Sorry! No results because I celebrated Christmas?"

"I'll help you from Boxing Day on, I promise! In addition, we have more experience with colonies on a comb with brood than with swarms!" "We'll help you, too!" two children's voices unexpectedly sound behind them. Niko has to admit defeat and only drives to the villages to check the news.

The palm tree is beautifully decorated, with homemade baubles, angels, stars and colorful chains. Candles were easy to get. Since the electricity fails more often, there is probably no household in Anuradh that does not need candles. How to attach them to the palm leaves? Gudrun has a housewife idea: "We clamp them in clothespins and tie them to the branches."

Tying them down doesn't work, they are much too crooked.

"We need a heavy counterweight on a wire at the bottom. That keeps the candle vertical," Niko interrupts. "Why is that?" asks Fritz immediately. "Think of the stand-up puppet you loved so much. They have a lot of weight at the bottom, and no matter how hard you push them, they always stand vertically."

"A stand-up candle! Great, we'll start tinkering right away." "Do you have any idea how to do that?" "We had candlesticks at home that had a short wire under the candle, and at the very bottom we hung a heavy, metal pine cone. We have thick wire, but what do we use as a counterweight? It has to be small and heavy and as inconspicuous as possible."

They run back to the only shopping street. They are looking for a shop with hardware. Fritz discovers a box with slightly rusted medium-sized nuts: "I've got it! We can hook the nuts at the bottom of the wire. Now our Christmas tree will be complete." The children are practically running home and are doing handicrafts when Niko and Gudrun arrive. And it works, the candles are straight!

Fig. 35. This Areca palm has to serve as a German Christmas tree.

Indrani has outdone herself! A table is set up with lots of rice and five different curries. The cakes are on another table, Gudrun has bought the bakery almost empty. Guests have to bring plates and spoons, because parties had not been planned when the house was furnished. Tilla and his family are of course there, the German students along with Punchi and Noel from Kandy, the taxi driver Chandaratne from Kandy, Mr. Ossen and Mr. Sellaperuma, the "assistant Bee Catcher" and Mr. Samarakoon. Indrani and Gudrun have to pay attention to the food. As a welcome, there is arrack and fruit juices. The arrack is drunk neat or in a cocktail with fruit juices. There is a lot of drinking before the guests start eating. The rice and curries get cold.

Gudrun quietly asks Tilla why no one eats anything. Tilla has to laugh: "When you drink arrack after dinner, you hardly feel it. And besides, the curries have to cool down anyway. We eat with our hands, otherwise you'll burn your fingers!"

Finally, the candles are lit on the decorated palm tree and the children sing Christmas carols with Gudrun and Niko. That has to be done.

"A true German Christmas with a real Christmas Tree!"

That's even true, almost! The golden balls made from the wrapping paper of the Gold Leaf cigarette glitter in the candlelight, as do the silver angels made of aluminum foil. The colorful paper chains also come into their own and the candles stand "straight as a bolt". There is a small gift for each guest and the children soon disappear into Tilla's garden. Not for long, because they quickly come back shouting loudly: "We've discovered something. In the front garden, a swarm of bees hangs from the tree. It's really big."

Before they finish talking, Niko has rushed out of the room like lightning. Tilla, Mr. Sellaperuma and Gudrun also run outside. And really, there is a beautiful large swarm of the giant honey bees hanging there. "Now I just wish that it finds its final nesting place here nearby. Then the research can begin." A happy Niko goes back to the celebration and there is a toast to this special Christmas present.

Thirty-seventh chapter

Niko and Noel observe drone flight

After the arrival of *Apis dorsata,* Noel wants to start his project as soon as possible and asks: "We want to observe the flight times of the drones in the different bee species. What is the scientific benefit of this work?"

"We want to understand how the three bee species in Sri Lanka co-exist. How do they manage to live so well together," explains Niko. "The avoidance of mating across the different honey bee species plays a major role in this. It was only about 15 years ago that the queens' sex pheromone, which attracts drones, was chemically analyzed in the Western honey bee. Shortly afterwards, the queens of the three Asian species were also examined, but contrary to expectations, in all three species the sex pheromone is similar to that of our western queen bee."

"I wouldn't have thought that. During our studies, we learned that sex pheromones are different in different species!"

"The honey bees must have developed a different solution here. In Oberursel, at any rate, the natural pairing went wrong. *Cerana* queens and *mellifera* drones were in the same place at the same time. Mating occurred, but the *mellifera* drones could not separate from *cerana* queens. Both fell to the ground and the ants were happy about the food! Reproductive isolation is the basis for closely related species to survive together in the same habitat. If all the drones are flying at the same time, could the drones visit different places specific to each species?"

"Yes, that could be the case. Even an equal drone flight period for all species would be a new scientific result. At the moment, nothing is known about it. Enough discussion. Let's get started! For dwarf honey bees, I will look at *Apis cerana*. We have enough colonies with drones for these species. We still have to wait for more colonies of the giant honey bees."

The next morning at six o'clock, both sit in front of their colonies with a stopwatch and protocol booklet. Towards evening, they compare the flight times: The drones of *Apis florea* fly in the early afternoon, while *Apis cerana* drones only fly after 4 p.m., when the last drones of *Apis florea* have stopped their mating flights.

"Rarely have I seen such clear results!" exclaims Niko enthusiastically. "Clearly different times in the mating flight! No one has had this idea before!!" Noel is also happy, but expresses it more quietly: "Obviously you expected that. Let's repeat this with other colonies and on other days." "That's right, Noel, we have to log many days and at least five colonies per bee species. We will manage to do that in the next few days. Nevertheless, we can toast today's results, they are so clear!" Niko's usual optimism is unmistakable. This is followed by a cheerful evening in a large group (Fig. 36).

They did not expect it to be easy to observe drones with *Apis dorsata* colonies. Mr. Sellaperuma reports that a large *dorsata* colony had been found at Mihintale, a very old Buddhist temple. Niko drives off immediately and wants to see where and how it hangs. Mr. Telambura, who delivered the message, refuses to show them: "These bees have stung some people. We give this square a wide berth." Niko takes this warning seriously. Mr. Sellaperuma and he put on protective suits with veils, gloves and boots. Even before they reach the nest, they are surrounded by *Apis dorsata* guard bees. The first bee stings in Niko's glove. From the experiments during his previous stay in Kandy, Niko knows what will inevitably follow. The sting in the leather of the glove releases alarm pheromone and attracts many other guard bees. After a few seconds, he can hardly see anything through the veil. Too many bees sit on the wire, bite and try to sting. "Where to?" They can neither go back to the car nor to the settlement. The people there have no protective clothing, the stinging bees would cause panic. Mr. Sellaperuma points to the east and walks slowly. Niko also follows slowly and carefully. Under all circumstances, they must prevent thorns or undergrowth from tearing their bee protection. The number of attacking bees does not decrease at first. The first guards have overcome protective clothing. Mr. Sellaperuma groans. Several bees have stung him in the face. Niko is better off. His bee veil is still dense. Only above the right boot is he stung several times. Cursing and moaning, they both walk on. The distance from the bee colony increases and finally the number of stinging pursuers decrease. After half an hour, there are only a dozen bees left. Niko is exhausted and sits down on a tree trunk. Mr. Sellaperuma also stops. They retreat to the shade of a tree and get rid of their sweat-soaked protective suits.

"Sellaperuma, are you all right?" "Yes, sir, all good. I go and fetch Mr. Ossen and the car. You'd better wait here until I'm back." "It's okay. Hopefully it won't take too long! I could use a sip of water!"

An hour later the car arrives and shortly afterwards they can relax with a cup of tea at Mr. Telambura's. After dark, Niko and Mr. Sellaperuma go back to the giant honey bees. In the dark, the bees cannot orient themselves. With the help of the flashlights, they finally find a large *Apis dorsata* colony. They carefully remove some smaller tendrils and branches that obstruct the view of the comb. Then they cut a small space about two meters in front of the comb and set up a chair they brought with them. Finally, they can drive home. A long, exhausting and painful day is coming to an end!

The next morning, the alarm clock rings at 4 a.m. It is still pitch dark when Niko makes his way to the bee colony with protective clothing, water bottles and cream crackers. With first light he experiences a bad surprise. The *Apis dorsata* comb shows clear alarm reactions. The guard bees form long chains at the bottom of the comb and the first bees take off in a defensive flight. Immediately and in a great hurry, he puts on his protective clothing and escapes. Just in time.

Noel, who arrives back in Anuradh from Kandy the following day, is disappointed when he hears about Niko's failed experiments to observe the drone flight at *Apis dorsata*. "What can we do? Is there no protective equipment that can withstand the attacks of the giant honey bees? Niko, do you want to give up?"

"No, I never give up. I heard from Professor Morse years ago that he had achieved almost perfect protection against *dorsata* stingers with the protective suits of the forest firefighters in Montana. We don't have these suits. We have to find another solution." Gudrun is joining them. "Gudrun, what can we do to protect ourselves against *Apis dorsata* and watch our drones?" "You have to prevent alarm pheromone from being released. Without the alarm pheromone, there will be no targeted attacks on you." Niko looks surprised. He wants to answer. Then he sees that Noel probably also has a response: "Gudrun, that's a good idea. We have to prevent the bees from stinging!" Noel is probably not familiar enough with the colony defense of *Apis dorsata*. Niko does not want to offend him at all: "Noel, we can't prevent guard bees from taking off on stinging flights when disturbances occur, and bee researchers in front of the comb cause massive disturbances for many colonies."

Noel mumbles softly. Gudrun and Niko don't understand him.

"Noel, a little louder, please," says Gudrun with a smile. "Please, not as loud as Niko." "We cannot prevent flights with the intention of stinging. That's clear. In order to raise the alarm, the bee sting must penetrate the skin or even the fabric to get stuck there with its barbs. This is the only way the bee can mark an enemy. This obviously works very well with your protective suits." "You're right, the firm, thick fabric of our overalls, the leather gloves, the hat and the seams of the bee veil provide enough support for countless bee stings." "Could we convert the protective suits? "Niko shakes his head and turns away.

"Wait a minute, Niko. A protective suit made of material resistant to bee stings is not a solution for you. Think of the Sigiriya rock with the cages. They would be too big, a small cage made of wire and wood in front of the comb would not offer the bees the opportunity to anchor their stingers," Gudrun interjects. "You sit in the cage and watch your drones. In the cage you are safe: the bees cannot reach you there and, above all, cannot use their stings and release the alarm pheromones!"

"Yes, that could work. I'll make a construction drawing right away and Charly, our carpenter, could build this cage immediately."

The observation cage is ready the next evening. It is brought to Mihintale on the roof of the Peugeot and installed in front of the comb at night. Even before daybreak, Niko is sitting in his protective cage (Fig.37. Fig.38).

Fig. 36. Noel Wijeyagunesekara observes drone flight of an *Apis florea* colony

The bee colony initially reacts with defensive reactions, and some guard bees react, flying. They circle the cage and land on the wire and the white, thin and taut cotton fabric. However, all stinging attempts are unsuccessful, as planned. No stinger sticks. Thus, the alarm behavior subsides significantly after only half an hour and dances increasingly take place on the comb. Many of the returning bees carry pollen pellets on their hind legs. Around noon, the signs of defensive behavior disappear. Niko cautiously ventures a first short excursion behind the cage. Immediately, restlessness arises on the comb. He quickly retreats to his cage. The afternoon heat is getting to Niko. Even waiting in vain for the drones to take off does not contribute to his well-being. Around 5 p.m., his water bottles are empty. He breaks off the observation and runs, undisturbed by the bees, to Mr. Telambura's house. There is hot tea! Niko waits for Mr. Ossen, who arrives at 6 p.m. as agreed.

In Anuradh, Niko reports on his observations: "Our sting-resistant protective cage has proven to be excellent. Noel, you're right, without alarm pheromones, a *dorsata* colony calms down relatively quickly. In the end, I was able to leave the cage and walk safely, without being attacked, to Telambura's house! However, I have not observed any departing drones. Tomorrow I will take a third bottle of water with me. It's very hot there and I need a lot of water to tolerate this heat."

"Do you want to sweat in front of this colony again from sunrise to sunset?" asks Gudrun. "Why do you think the drones will fly tomorrow?"

"Well, the weather and other circumstances may be more favorable tomorrow than today! The drones may also have gotten used to our protective cage!"

Niko doesn't sound convinced.

Noel listens in silence. He spent the whole day exploring the temple region and the ruins in the ancient city. Now he casually reports on his experiences:

"Tomorrow I will probably be able to observe a completely peaceful *dorsata* colony. Without a cage and at close range." "How is that?"

"At the Ruanceli Dagoba, about three meters high, a colony nests (Fig.39). Many hundreds of pilgrims and monks walk around the Dagoba every day, praying and walking close to the colony. The bees are absolutely peaceful and do not sting anyone. I spoke to the abbot. He and the monks think it is a sacred place that the bees respect. I have also been given permission to observe the bees as long as I respect the sacredness of the place. Of course, I agreed to that."

Noel smiles. Obviously, he enjoyed the tour through the district in the old city with its many temples and dagobas.

"Great, Noel, Mr. Ossen will drop you off at the Dagoba tomorrow morning at 5 o'clock. I will drive on to my colony in Mihintale. At some point we will find out when the drones fly."

The next day, Mr. Ossen arrives at Mr. Telambura's house around 7 p.m., Niko has been waiting for a long time. "Sorry sir, a puncture!" Mr. Ossen had to mount the spare wheel."

"Mr. Ossen, we should hurry. Noel is also waiting." Mr. Ossen accelerates for once and gets the old Peugeot going. On the way to the Ruanceli Dagoba, Noel meets them. "Sorry, we're late. There was a flat tire," Niko apologizes.

"No problem," says Noel cheerfully. "By the way, I solved our problem. The *dorsata* drones on the Dagoba take off at dusk and fly only very briefly. At 6:10 p.m., all the drones are back and land on the comb."

Niko is speechless. Yes, he always stopped observing at the beginning of dusk and set out for the house of Telambura, exhausted, thirsty and tired.

"Noel, congratulations, you have the necessary stamina required for scientific work. I myself gave up too early, I thought that honey bees would not fly at nightfall and so missed the drone flight! According to your results, I will observe from now on at least until 7 p.m.," assures Niko. "Well, I only sat with the colony so long because of your lateness," Noel explains modestly. "You're right, we have enough observations from 6:00 a.m. to 5:00 p.m. Now we'll just watch later until dark!"

No sooner said than done, and with success! The drone flight periods of the *Apis* species in Sri Lanka are well separated in time. *Apis florea* drones fly approximately from 2 p.m. to 3 p.m., *Apis cerana* from 4 p.m. to 5:15 p.m. and *Apis dorsata* from 5:40 p.m. to 6:10 p.m. The corresponding publication by Niko and Noel will one day be one of the most cited scientific papers on Asian honey bees. However, many people complain about Noel's last name "Wijeyagunesekara". Noel counts more than 10 different spellings of his name in the quotes by various authors.

Thirty-eighth chapter

Gudrun and the students offer Lego bricks to the bees

Except for the experiments with the giant honey bee, all others are carried out in Anuradhapura. Punchi and Micha train the three species to feeding places at different intervals. Bees are marked at the feeding ground and their dance speed is logged per distance, while Conny and Gudrun concentrate on the dwarf honey bees. When these

Fig. 37. Transport of "sting-proof" protective cage.

colonies are transported to an unfavorable place, they soon leave the comb. They observe how these bees coordinate their move.

Gudrun is often busy with the children. In addition to mathematics and English lessons, German, geography and social studies also must be taught, which is Gudrun's job. Twice a week they go to the swimming pool in the Resthouse. Janthe and Kantschi are allowed to come along, according to Sita and Tilla's wishes, so that they could learn to swim. Gudrun admires this decision, because at birth, the children were predicted that they would one day drown. In Sri Lanka, belief in such predictions is widespread. Even the prime minister, it is reported, sets all dates according to the horoscope. The Tilla family, on the other hand, knows that children who can swim do not drown. The situation is favorable, the children often have the swimming pool to themselves to play in and the two brothers don't even notice that they can swim after a few weeks. Tilla and Sita came along one day to see it with their own eyes. They are happy!

Anna and Fritz often occupy themselves alone. There is the dog with her two puppies which need to be cared for and who are always ready to play. Little Slender-Lori, who was given to them by neighbors and who lives in the palm tree in the living room, needs grasshoppers and other insects every day, which are caught with nets in the overgrown garden. They see snakes and once a large iguana comes running through the bushes towards the house with a roar, a moment of shock for the adults as well. Most of the time, however, they play with the Tilla children. This gives Gudrun time to remain active in research.

In the evenings, everyone often eats together at the large table, Gudrun and Niko with the children, Punchi, Noel, Sellaperuma and the two Frankfurt students Conny and Micha. Indrani obviously enjoys cooking for so many. Always very excellent! The conversation usually revolves around the experiments, but also often about small unexpected experiences. At one point, the shopping facilities in Anuradh were discussed. Conny is outraged: "There is not even Coca-Cola. The answer is always 'Out of stock!', then the shopkeeper smiles and says Coke will come next week."

Punchi looks surprised: "Conny, why on earth do you need Coke? We have tea and sugar. That's more than many people here can afford. And it can turn worse. I have heard that in the last few weeks ships with wheat from Canada have not arrived. Soon there

 Fig. 38. Protective cage in front of an *Apis dorsata* colony.

will probably be no more bread."

"These are bad prospects." Micha sounds worried. "Then we won't have enough to eat?" Punchi laughs: "We have rice and corn. The harvest is good this year. So, dear ones, none of you will go hungry! However, the situation is different for many poor people. Every bottleneck in the supply here immediately leads to rising prices."

"Punchi, you're exaggerating again! Mrs. Bandaranaike, our prime minister, will make sure that there is no major famine," Noel intervenes. If Punchi's announcement comes true—there will be no bread for four weeks!

Then Mr. Sellaperuma speaks up. "Conny, I can get you Coke. Yesterday at Cheapside, you know the shop in the main street, I saw a crate of Coke bottles in the back of the shed. The shop belongs to a Tamil. He has close ties to Jaffna and there are smuggled goods from India. Conny, how many bottles of Coke do you want?"

"That must not happen," Niko intervenes. "German bee researchers buy contraband! We cannot afford such a headline in the local press." He turns to Conny and Micha:

"We are guests here in Sri Lanka. We respect laws and local rules. No smuggled Coke in our house!" "It's a pity," Anna interjects. "We haven't drunk a Coke since we've been here." Fritz, on the other hand, asks: "If there is no bread, do we get corn on the cob to gnaw off every day? I love it!" "Enough now," Gudrun shouts. "It comes as it comes! What is certain is that you two have to go to bed now."

When Gudrun returns, the experiments are being discussed again.

"Noel and I have completed the work on drone flight times. A perfect reproductive isolation of the *Apis* species by dividing the drone flights over time. Time for a new project. We should start with the experiments on feeding competition."

"What do you have planned?" Micha: "Might we help?" "Yes, it's about creating a competitive situation between the three bee species. We need all of you. We need to train bees from one colony of each species on an artificial food source. For each species, we need a feeder and a caretaker. We must make sure that only bees of one species collect the food. Bees of other species will be caught and caged. Then we shift with the food towards a final competitor site, where all species collect at the same time in the same place. About three meters in front of the competitor area, all previous feeding

Fig. 39. *Apis dorsata* colony at the Ruanceli Dagoba. Three meters distance to passing and praying monks!

containers are removed and a single new feeding container is set up in the competition place. Here we must count how many bees of each species collect food. In the end, as Gudrun's preliminary experiments have shown, one species usually succeeds in displacing and excluding the competing species. Gudrun, do you want to add anything else?"

"You have said the essentials. I will explain many details to you on site during the experiment. By the way: We use Lego bricks as feeding containers (Fig.40). The bees can suck there without smearing themselves with the sugar water. We will meet tomorrow morning at 10 a.m. in Tilla's garden. We start with the three *Apis* species. Punchi can take over *dorsata*. Micha takes *cerana* and you, Conny, *florea*. Agreed?"

The three nod in agreement.

The experiments prove to be difficult. Micha, in particular, often fails to prevent an *Apis florea* from successfully collecting sugar water on his Lego, which is intended for *cerana*. As a result, *florea* recruits new collectors and they compete with *cerana* far too early. Gudrun, who supervises the progress of the experiments, is crushed on the evening of the first day of the experiment: "Niko, you can't imagine how clumsy Micha and Conny are. We have not succeeded in successfully starting even a single experiment with the three species. Can you help tomorrow?" "No, unfortunately not, tomorrow I have to watch a swarm of migrating *dorsata* in Galkulama with Mr. Sellaperuma. The bees showed their first migration dances today. Tomorrow the swarm is supposed to start. I have to record that. In the future I want to collect the *dorsata* swarms and set them up here in the garden. Then I can also help with the Lego tests." Gudrun complains: "Niko, that's not fair! You can't plan so many experiments in parallel, especially if you employ inexperienced students as helpers. They need your supervision. I can't put unwanted bees in cages at my feeding station and help students who are sitting 100 meters away in different places at the same time. Couldn't at least Punchi and Noel help? At least tomorrow, the two are experienced in experimenting with bees!"

The Lego tests went better in the next few days. Slowly the results are emerging: the small *Apis florea* often remains the winner and eliminates *Apis cerana* and *Apis dorsata* from the sugar water in the Lego brick. In the evening round there will be discussions. "For me, it's contradictory. The large *dorsata* worker is more than three times as heavy as the small *florea*! If both fight, the great *dorsata* should win," says Conny. "Conny, there's no doubt about it: the experiments show that little *florea* wins more often. My calculations statistically show a very clear dominance," argues Gudrun. "Well, maybe it's not worth fighting for *dorsata*," Punchi points out. "My experiments show that *dorsata* has a very large flight radius. They can easily switch to other flowers and don't have to waste energy displacing competitors. *Florea*, on the other hand, has only a small flying range. It is more worthwhile to defend a good source of food once discovered." He looks around. Obviously, he is waiting for disagreement.

"Punchi, that sounds logical. We must keep in mind that Lego bricks filled with sugar water are something very artificial. For me, the transfer of our Lego test results to natural collecting behavior is still questionable," Noel states, shaking his head. "I watched the bees visiting flowers and never saw that there was a fight between different species. In most cases, bees only land on the flower when they are sure that no other bee is currently collecting there. Shouldn't we first show that there is an overlap of food sources between species?"

Niko takes the floor: "Noel, that's clear. We see again and again that good nectar or pollen plants are visited by all *Apis* species. By the way, we are expecting Dr. Günther Vorwohl, probably one of the most internationally renowned specialists in pollen analysis. We want to collect honey samples from the bee colonies in this area with him and will then gain precise knowledge of which plant species are visited by all bees, and whether there are any flowers at all that are only used by one honey bee species."

Thirty-nineth chapter

Gudrun and Niko have a visitor

Somehow, bee researchers seem to communicate with bush drums. One hears the tam-tam and passes it on. And it's "Tam-Tam" around the globe! One afternoon, Professor Jurek Woyke from Warsaw shows up at their door in Anuradhapura. Wet with sweat, a handkerchief knotted at the four corners over the head, his face reddened by the sun. Over his shoulder hangs a worn-out travel bag. All his luggage!

Gudrun and Niko have known Jurek Woyke for many years from their visits to Poland. "Jurek, how did you find us in Anuradhapura?" asks the puzzled Niko.

"Well, that was easy. I come from India, from the Bee Institute in Poona. They heard from Professor Ruttner that you are both in Sri Lanka at the moment. So, I flew from Trivandrum in South India to Colombo. I remembered Baptist and visited him in Peradeniya. From him I learned that you are here. The most difficult thing was the way from the train station to your house. Twice I was shown the wrong way."

Fig. 40. *Apis florea* and *Apis dorsata* compete for sugar water at a Lego brick.

"Jurek, come in," says Gudrun. "Do you want some tea? Please sit down!"

"Thank you, tea would be good. But, if possible, I would like to take a shower first. I have a towel with me."

Jurek takes his travel bag and Gudrun shows him the shower behind the kitchen. Later, they sit together over tea. Jurek wants to stay for a few days and is mainly interested in *Apis dorsata*. "You have brought *dorsata* colonies to Oberursel. Niko, I'd like to see you get the colony in the box and transport it. Would you show me that?"

"You're lucky, we're in the process of preparing a colony. Probably we'll get it in two days in the dark. This time only as far as here in the garden. You are cordially invited to be with us." "Gudrun, do you still have a room available for me? I'm happy to pay for it." "Jurek, sorry, we are totally booked. We have outsourced our Sri Lankan students to neighbors. Why don't you go with Noel to Mr. Tillakaratna and ask if he still knows of a available room. It would not be expensive with the advantage of being not far from us.

You are welcome here for dinner." "Thank you, Gudrun, very nice. If I have time, I would like to eat with you. Niko, please let me know when you go to *dorsata*."

The next morning, Niko and Mr. Sellaperuma take Jurek Woyke to the *dorsata* colony. "We now want to measure the comb, the branch and, above all, the exact angle of the branch. In doing so, we must not cause an alarm or a defense behavior. That would make the measurement impossible," says Niko.

"I understand that," Jurek answers. "Let me help." "No, we must wear veils and protective clothing! Please stay with Mr. Ossen at the car. You can get into the car and be safe if it comes to full defensive behavior!" "Niko, I had bees when you were still in kindergarten," Jurek replies angrily. "Jurek, this is not about life experience. I want, no, I have to measure this colony. I can't allow any risk to be taken. You have the choice: either you accept my decision or we will take you back to Anuradh!" Jurek remains silent, apparently offended.

Measuring is time-consuming because individual tendrils and lianas have to be carefully removed. Finally, all the measurements are noted down and they make their way back to the car. To their surprise, Jurek emerges from behind a bush. He has disregarded Niko's recommendation. "I wanted to see how you measure the colony. That's why I'm here." "Well, everything went well. Let me just say, Jurek, that you were frivolous. In the case of *dorsata*'s defensive behavior, you would have had great difficulty!" "Believe it or not, I can take care of myself and have mastered greater dangers," Jurek grins. Niko remains silent and resolves not to let this Professor jeopardize his plans in the future. He makes the construction sketch for a collection box and brings the drawing to the carpenter Charly: "When can you have it ready?" "How fast does that have to be? I have to buy jute and strips. Everything else I have in the workshop." It is not the first *dorsata* box and Niko knows that he can rely on Charly. "The day after tomorrow in the afternoon? Can you do it? And can you help us at night to secure the box with the bee colony for transport?"

"It's okay." Charly nods several times.

At around 5 p.m., Charly arrives, as promised. Professor Woyke has also come and inspects the extensive equipment together with Niko. Mr. Ossen compares piece by piece with a list he got from Niko. "Everything is complete. Only the lamps are missing!" Mr. Wilson also arrives, from whom Niko has rented an old Land Rover for the night. Mr. Sellaperuma and Niko get into the Land Rover. The others go with Mr. Ossen. They go to the crossroads in the direction of Mihintale to Mr. Telambura's house. That is the assembly point and at the same time the end point for Mr. Ossen and the Peugeot. Niko explains the plan of the operation: "Sellaperuma and I continue with Wilson to the colony. If Professor Woyke wants, he can also come along. Mr. Wilson will then drive back here to the assembly point. Mr. Sellaperuma and I will hang the box on the branch and seal everything. Then we saw off the branch and put the box on the ground. Finally, Mr. Wilson arrives and we bring the box with the bee colony to the assembly point. In the end, we'll all meet here again!" (Fig.41).

Jurek sits down next to Niko in the Land Rover: "Jurek, don't you have the veil we gave you?" asks Niko, in full gear. "No, I don't need protection. I will hide silently and motionless in a bush. The bees won't discover me and I can watch how Sellaperuma and you get the bee colony into the box." Niko is surprised at so much arrogance. Apparently, Professor Jurek does not accept his advice. "As you like. Hopefully you won't regret this decision."

Niko and Sellaperuma carefully carry the box under the branch on which the colony is hanging. Mr. Sellaperuma lifts it, and the shaking of the branch triggers a full defense. Sellaperuma and Niko are covered by stinging bees. Niko attaches the box with the prepared wires to the branch. Now the branch carries the box. Sellaperuma and Niko pull the jute over the branch and seal the side openings. Now it is a matter of waiting until the stinging flights subside. At the same time, Niko carefully begins to saw off the outer end of the branch close to the box. Then he moves under the box, which he holds with both arms. Sellaperuma saws off the other end of the branch and Niko carefully places the box on the ground. Sellaperuma now fetches Wilson, who is frightened, his face protected by a bee veil, and he helps lift the box into the Land Rover. In the meantime, many residents have also gathered at the assembly point. They like to take the cigarettes of the "Gold Leaf" brand offered by Niko all around. Mr. Telambura arrives, to whom Mr. Sellaperuma hands over 100 rupees for the bee colony as agreed. Charly, smoking his second cigarette, takes care of the box with the bee colony. He secures the jute cover with strips and nails against possible leaks.

Suddenly, Sellaperuma shouts: "We have forgotten the Polish Professor! Should I check?" Sellaperuma starts running. After half an hour, he comes back alone. He didn't find Professor Woyke. Niko has to bring the bees to Anuradh and set them up in the flight cage in Tilla's garden as soon as possible. What should he do? Mr. Telambura offers to send his "boys" to look for Jurek. "Thank you very much, that's good. We are now bringing the bees to Anuradh. I'll send Mr. Ossen back. Can the Professor wait with you until then?" Mr. Telambura agrees. Niko and his people drive with the bees in the box to Anuradh and hang the bee colony in a cage on the prepared rack. At home, Gudrun is waiting. She is indignant that Jurek has disappeared. "You should have taken better care of Jurek. Hopefully nothing serious happened to him. He doesn't know anything about *dorsata*." "Gudrun, you say it, that's the problem. Jurek is arrogant and doesn't understand that the bees are more defensive here." Mr. Ossen honks in front of the house. Jurek sits next to him. His face is swollen. The lower lip hangs down. He looks miserable.

"Thank God, you're back! Come in. Do you want a drink?" asks Gudrun relieved.

"No, I've had tea at Mr. Tellambura's. I'm dog-tired and want to sleep. Tomorrow I'll be fine again." Jurek speaks stuttering and slowly, probably because of the swelling in the mouth area. "Jurek, what happened?" asks Niko.

"At first, the bees only stung Sellaperuma and you. Then I had the first stinger in my face. More and more bees came. I ran away, probably many kilometers, until I couldn't take it anymore. I sat down under a tree and there a man of Mr. Telambura found me. Could Mr. Ossen take me to my room?" "You're lucky, Niko," says Gudrun when the professor is gone. "If Jurek wasn't such a healthy, resilient guy, this could have gone wrong. You have put him in serious danger. That's not okay. After all, Jurek is our guest here." "You're right. How am I supposed to deal with a senior professor, who doesn't want to listen to a small scientific assistant? Let's try to get rid of this guest as soon as possible."

Fortunately, Jurek travels back to Colombo the next day.

Only a few weeks later, a servant of Professor Baptist's niece in Anuradh, Mrs. Mahadiulwewa, stands in front of the house. He delivers a letter announcing Baptist's visit with a Canadian Professor, Gordon Townsend, in two days. Niko has just returned from a strenuous tour. "Do you know a Professor Gordon Townsend?" asks Gudrun. "Do you mean Professor Townsend from Guelph in Ontario?" Niko asks back. "Yes, that will probably be the right one. There are certainly no two bee researchers with this name.

What kind of guy is he?" "Gudrun, you met Gordon not long ago at the IBRA conference in London. That was the one we talked about honey production in Asia and Africa. Gordon works for CIDA, the Canadian Development Aid Organization. He asked us about Sri Lanka at the time!" "Yes, now I remember dimly, as you know, I don't have a good memory for names. What does the Canadian want from us? Espionage?" "No, Gordon is not Jurek. No competition. We will welcome Baptist and him in a friendly and as good a way as possible. We show everything they want to see. Baptist in particular will be interested in how his students work here. Anuradh, with its mosquitoes, stinking water and poor hygiene, will rather scare off the cultured Canadian!" Niko grins and hugs Gudrun: "Gordon won't understand why we and our children like it so much here. Up to thirteen nice toads every morning in our shower, that's not Gordon's world."

As announced, two days later a large American Ford stops in front of the house. Professor Baptist and Professor Townsend get out. "Good Day!" Baptist beams: "This is my friend Gordon, you met him in London. We look forward to meeting you here in Anuradhapura and hope you allow us to take a quick look at the bee research workshop!" Gudrun overcomes her North German reserve and beams as best she can: "What a great joy! Welcome! Please come in. No, you don't have to take off your shoes! Please take a seat!" Gudrun calls for Niko and the children. After the "Shake hands", the children drink cold lemonade and the adults tea. Soon Janthe and Kantschi, Tilla's sons, come and run into the garden with Anna and Fritz. Gordon reports on his latest results from Kenya: "The topbar hives have proven to be very successful. There is a lot of honey. After many difficulties, first successes and recognition." "This is good news. My congratulations! Here, we are primarily researching the biology of the three honey bee species. May I invite you to a short tour. We are currently working on various topics. Noel observes the drone flight periods of the three honey bee species. Micha and Conny study food competition in the three bee species and Punchi is sitting upstairs watching the dances of *Apis florea*. In our garden and in the garden of our neighbor, Mr. Tillakaratna, we have set up large flight cages with *Apis dorsata* colonies!"

"Before you go, please give us a short update on the further plans," Gudrun asks.

"Sorry, Gudrun, unfortunately I don't have much time. Tomorrow there will be an important meeting with the Department of Agriculture in Kandy. I have to drive back this afternoon. And before I forget, we have an appointment with Mahadiulwewas at the Miridia Hotel for lunch around 2 p.m.!" "Oh, No! So, you must come again!"

During the inspection tour, Niko holds back emphatically and asks the students to report on their experiments. Gordon is surprised at the calmness and serenity with which these free-living, supposedly wild bees are experimented with. His questions about bee veils, gloves and smoke surprise all students. Noel with the giant honey bees also wears no protection and explains: "In the meantime, I know the behavior of the bees quite well and I can observe the colony well from a distance of 1 meter. Only when we have to do something directly to the colony do we need protective clothing."

"Apparently, African honey bees are more defensive than their Asian sisters," Gordon comments. During the short farewell, Gudrun and Niko once again thank Professor Baptist for the help of his two students. "The mix of students from Sri Lanka and Germany helped us a lot. The knowledge of local conditions has contributed decisively not only to the success of our work, also to the understanding of the way of life in Sri Lanka. And Gordon, as Gudrun has said, you must come again soon! Have a good journey!"

Finally, Professor Vorwohl arrives, exceptionally a guest with prior registration. He has reserved a room at the Miridia Hotel and is full of energy. Niko is on the road with him

Fig. 41. *Apis dorsata* colony in a transport box. The front was opened after the final experimental location was reached.

during the following three days to collect as many honey samples as possible from the three honey bee species. Ideally, the colonies should nest as close together as possible. Niko has taken precautions and knows where it is worthwhile to collect. Fortunately, this guest is sensible, he lets Niko and Mr. Sellaperuma work on the bees in peace and watches them from the car, protected from the bees. Then he flies back to Germany, happy and satisfied with many honey samples for pollen analysis.

Fortieth chapter

Gudrun and Niko drive to the sea and catch the return flight at the last minute

"Niko, we've been here for four months now and you promised the children and me a week at the sea." "What you promise, you have to keep!", Anna intervenes. "All the other children have been by the sea, except Fritz and I."

"Anna, we had more to do than planned. You're right, a promise is a promise. We will go to the sea." "Gudrun, what do you think?" Niko takes up the topic again in the evening. "We pack and break down our tents a week before the return flight. We'll go to the sea and then get on the plane!" Gudrun groans: "Typical Niko, the well-deserved vacation only at the last moment. You always have new ideas about what you still want to do. Looking at you like that, you'd rather spend a week longer with your bees than with the children and me at the sea." Gudrun gets up and retreats to the bedroom. Niko takes out his large diary and enters in red pencil for March 16: "Departure to the East Coast!" Then he goes back in the calendar. From packing the suitcases to the program for the next day. It's not that simple. After many changes, he finally drew up a work plan that probably contains most of the essential things.

"Gudrun, we're going to Trincomalee on March 16th! Okay? Ask the Mahadiulwewas which hotel on the beach is good for us and the children? We should reserve a bungalow as soon as possible." He adds: "I'm looking forward to the sea with the children and you!" "At first only bees in the head, and suddenly the loving father and husband. I wonder how much longer I'm going to put up with this?" Gudrun now sounds conciliatory.

The day before their departure on the 16th, Gudrun and Niko said goodbye to everyone: Indrani gets an additional month's salary. Gudrun got her a job at with German family in Kandy. Sellaperuma was delighted with the new bike—"Flying Pigeon", made in China. Mr. Samarakoon strokes the two young dairy goats he had wanted for a long time. All this with the firm promise that there should be no elaborate farewell the next morning. There they are now: Indrani with tears in her eyes, a doll for Anna and a car made of tin cans for Fritz. Mr. Sellaperuma hands over an elaborately carved ebony letter opener, which Charly must have made. Mr. Samarakoon hands Gudrun a bag of warm string hoppers and mumbles something like "Sri Lanka Breakfast". Janthe and Kantschi, Anna and Fritz's playmates, arrive. Niko is touched. He wants to say something, his voice fails. Gudrun steps in and underlines the harmony, the help and the affection they have experienced. Then they hear the horn of the old Peugeot. Mr. Ossen wants to be at the Nilaveli Beach Hotel before dark. Now everything is happening very quickly. Gudrun and the children sit down in the back seat, Niko in the front next to Mr. Ossen, they wave and off they drive.

They take a break at Polunaruwa Rest House. Tea and cola brought along for the children. Then the heat takes its toll. Niko and the children fall asleep. Only Gudrun looks out of the window and admires the vegetation, which is more diverse here than in dry Anuradhapura. Suddenly she feels a blow, followed by rumbling. The car skids before it comes to a stop at the side of the road. Niko knows what's going on and calms down the children, who have woken up frightened. No houses in sight, only dense tropical dry forest. "Don't worry, we only have one flat tire. Probably a thorn in the tire. That's not a bad thing. We have a spare wheel," Niko reassures. Mr. Ossen points to the right rear wheel. The car is on the rim. First the luggage has to be taken out of the trunk and then the spare wheel and the jack.

"Where are we?" asks Fritz sleepily. "On the way to the sea, probably about 40 kilometers from Trincomalee," Niko answers. Then, seemingly out of nowhere, some residents appear between the trees. One of them carries a chair and puts it in front of Gudrun: "Madam, please sit down." "We stop at the road in the middle of nowhere and there are people there to help. That only exists in Sri Lanka," Niko notes. Two young men help Mr. Ossen change the wheel. "Sir, the spare wheel is mounted. We can continue," he reports. Niko checks the tire. "A completely worn tire. No profile at all. Will we get to Nilaveli in one piece?" "Yes, sir, we're driving slowly. It will be fine," assures Mr. Ossen. "We are in the hotel for five days. That's enough time to repair and install the leaking tire in a workshop in Trincomalee." "It's done, sir!" Mr. Ossen confirms with a hint of a bow.

When they arrive at the hotel, Gudrun cannot find the reservation number. They are supposed to wait for a "shift manager". The children are thirsty and hungry. "We go to the buffet and eat there. If your boss comes, he can find us there." Gudrun doesn't allow any debate. The large luggage remains in front of the reception, and they have found a free table. Niko remains seated at first. Gudrun goes to the buffet with both children. It takes quite a while for them to be back. Anna's plate is heavily loaded with pudding and cake. Fritz has a very large portion of French fries on his plate. Gudrun's plate reflects the rich buffet and the nearby sea. A waiter brings lemonade and a large bottle of water. Then an older gentleman comes to the table, who greets them in a friendly manner and does not skimp on the words "Welcome" and "Sorry" in many variations. Niko follows the manager to the reception and fills out the registration form. They go to the bungalow with two porters. It takes half an hour for Niko to return to Gudrun's table with the room keys.

"Thanks Niko! Get something to eat. You must be terribly hungry!"

Fritz and Anna accompany Niko to the buffet and have the dishes explained to them. They don't want chicken or fish. They do not refuse a large portion of ice cream.

The next morning, Gudrun hears the voices of the children and quickly slips into her swimsuit and dressing gown: "Come on, let's go swimming!" "Anna, I'm going with Fritz in the water. Come with me for a swim!" Gudrun makes a "belly flop" and gets the children wet. Anna grumbles and Fritz complains too. Then there is a fierce water fight. Finally, Gudrun gives up: "Mercy, Mercy! I can't take it anymore. What do you think of breakfast? First in the shower. We have to rinse off the salty seawater."

Niko is still asleep. "Niko has worked too much the last few days. He needs sleep," Gudrun whispers, grabs the children and goes to breakfast with them. When Niko wakes up at ten o'clock, there is a note on Gudrun's bed: "We're on the beach!" He grabs a cup of tea and three slices of toast in the lobby and starts searching. Even from afar he sees a large sand castle:

"Hello and good morning! You didn't wake me up? What did I miss?" he asks.

"We were in the sea and did a splash fight. Gudrun has lost! "I'm hot. I want to get into the water. Who's coming along?" Gudrun and Niko run off and the children follow.

Time flies quickly at Nilaveli Beach Hotel. Gudrun and Niko enjoy the warm sea and swim with the children for many hours every day. Boat trips to Pigeon Island with diving goggles, snorkels and fins allow new insights into a fascinating world of coral reefs. A boat trip to the nearby mangrove forests also shows the biodiversity of this habitat. Then everything has to be packed. On the morning of March 24, the Peugeot with Mr. Ossen is at the door. A quick breakfast and then they drive off.

"The day after tomorrow we will be back in Germany. Are you happy?"

Fritz answers: "Yes, I'm happy. At home, everyone speaks German. Everyone understands what I want to tell them." "Yes, I also want to tell everyone at home what we did in Sri Lanka," Anna agrees. Gudrun and the children look out of the windows. They have left the coast at Trincomalee, leaving the fields and the agricultural region behind. This is where the jungle areas begin, which is typical for the area around Polunaruwa. Niko leaves Sri Lanka behind in his thoughts. He thinks about what awaits him at home. How will the children find their way back to the German school after five months in Sri Lanka? Gudrun also worries Niko. Is this really only—as Gudrun claims—due to the overload caused by the children? In Sri Lanka, Gudrun has found her way back to harmony and her balance. At some point, Niko falls asleep. Then a violent push. The car skidded and then came to a standstill. "Mr. Ossen, another flat tire? Have you had the tire repaired?"

The person addressed smiles sheepishly. That, too, is an answer. "Anna, Fritz, Gudrun: get off. We have a flat tire again. This time it is worse. We don't have a spare tire anymore. Let's see how we progress!" Niko helps Mr. Ossen unload the luggage. Gudrun begins to get upset, again in the "German way": "Our plane leaves tonight. If we don't get to Katunaike Airport in time, they'll fly without us." "Gudrun, come down from the palm tree! I should have checked the repair of the tire in Nilaveli. I missed that. My fault! You can insult me. That doesn't change anything" Niko demonstrates his Sri Lankan smile.

In the meantime, again the first people come out of the jungle somewhere and talk to Mr. Ossen. Again, a young man comes with a chair. He also speaks English: "Please Madam, take a seat!" He pushes the chair to Gudrun. Gudrun has to laugh and notices how her excitement collapses. Yes, there is no point in getting upset. She thinks. "The people here all have charm and style. I'm sitting here on the side of the road in the middle of nowhere and I'm happy about this stupid chair!"

An older man with a bicycle drives at an unusually high speed for Sri Lanka and stops. Mr. Ossen, the flat tire on his arm, climbs onto the luggage rack and off they go in the direction of Trincomalee. That's probably about 30 kilometers, Niko estimates. Two hours by bike. Vulcanizing the tire half an hour, return by taxi another 30 minutes. "Our stay here is at least three hours!" Niko notes.

Swarms of mosquitoes attack them. Gudrun does not hesitate. She reaches into her hand luggage brings a bottle of repellent to light. Anna is the first to be carefully rubbed, then Fritz and Niko. Last it is her turn. "Gudrun, good to be stranded with you in the jungle. Without your "Autan", the mosquitoes would have eaten us! " Gudrun laughs: "Mosquitoes and a missed return flight. I wonder what can come now?"

Mr. Ossen reaches the airport in Katunaike just in time, or better almost in time. Thirty whole minutes to go. Now it's all about speed. A big tip to the porters. Some US dollars for the customs broker. They reach the gate only a few minutes before closing. "Niko, we agreed: No bribery." "You're right Gudrun, but a voluntary last-minute surcharge was inevitable. Or would you have preferred to miss the flight?" Gudrun mumbles something about label fraud. On the plane, they can finally relax. Niko and the children sleep first. Gudrun is awake even longer. She looks contentedly at her sleeping family before her eyes close as well.

Forty-first chapter

Niko works in Hohenheim and applies for a bee Professorship in Canada

In Oberursel, Niko evaluates the results of the Sri Lanka work and begins to write his publications. He has no problems publishing the prevention of mating between the three honey bee species (reproductive isolation) due to the different mating times. The detection of an additional alarm pheromone in the giant honey bee is also immediately accepted by a scientific journal. He supervises his doctoral students and takes care of teaching in zoology. He is overcome with numerous publications.

Gudrun is still involved in her research program on mating behavior, working all day in the summer months. From August onwards, she will be home at lunchtime and has time for the children.

The following spring, a permanent position as a research assistant at the Bee Institute in Hohenheim is advertised. At first, Gudrun and Niko don't talk about that at all. "It's not exactly my dream job," says Niko after a week. "The position would offer long-term economic security. The contract at the University of Frankfurt expires in two years. In Hohenheim, I can continue to work with bees and trust myself to remain active in research. Gudrun, what do you think? Should I apply?" "I also thought back and forth.

Here you have so much to do with teaching that you can only work on your publications in the evenings. Wouldn't you have more time for research in Hohenheim?" "Probably, but there I wouldn't have students any more. So far, we have only ever decided according to our research interests. We are becoming old and must not miss the right time to find a permanent position." Their elation during Sri Lanka with the exciting research and the great results has evaporated. Finally, Niko applies in Hohenheim. "I can still apply for other positions, possibly even for a professorship. I also habilitated in the old-fashioned way. I can apply anywhere with it!" He gets the research position and has to start in Hohenheim in September. Gudrun had to stay in Oberursel with the children. No apartment is found so quickly. She also wants to bring her research to a good conclusion. Her scholarship runs for another year. Gudrun remarks: "So far, we have always done everything together. Now a weekend marriage or weekend family!" "That's only for a short time," Niko assures the children and Gudrun.

After two weeks, when Niko—as usual—arrives in Oberursel for the weekend, Gudrun is standing expectantly at the front door. She holds a bee journal in her hand and calls out excitedly: "Niko, have you seen? The Journal of Apicultural Research advertises a professorship in Canada. They are looking for a professor who has specialized in tropical bees. You have to apply". "Let me come in first!"

Niko is tired and strained from the long motorway drive. "Do you have a cup of tea for me?" Of course, the teapot is well filled with aromatic Sri Lankan tea. "Do you remember the long discussions about your work with tropical bees? The unanimously negative verdict on the wrong choice of topic," adds Gudrun excitedly, "everyone agreed: a scientist specializing in tropical bees will never find employment! And now comes this advertisement. How many colleagues have research experience in the tropics? I know only one and his name is Niko! You have to apply quickly. This is your Professorship!" Niko reads the job advertisement carefully: "Yes, that sounds good! Of course, I want to apply. We should ask Professor Roger Morse of Cornell University for a letter of recommendation. He is likely to have some influence. Gudrun, as far as I can tell, there will be a lot of qualified applications. So much for my professorship, that's unlikely!" "Niko, maybe you're right. The best application falls by the wayside if local interests and those of the faculty are given priority. Canada is very far away. We don't know the conditions there. We have to try at all costs."

Niko asks about the plans for the weekend. Gudrun wants to use Saturday for her experiments. Niko agrees with Anna and Fritz to go to the indoor pool. "You can romp with us much better than Gudrun," says Fritz. "This time Anna and I will manage to submerge you for a really long time" Niko has just enough time to ask Gudrun: "Gudrun, should I cook something? Are you there for lunch?"

"No, I'm staying in biochemistry over lunchtime, probably I won't get home until around five o'clock. I prepared fried rice." Niko is disappointed. He had expected a Saturday with Gudrun. Gudrun senses the disappointment, but she can't change it. Everything has been complicated since Niko has been working in Hohenheim during the week. With the irregular school hours of the children in primary school, Gudrun usually only has a few hours in the morning for her research work. Her involvement in the BUND (confederation for Environment and nature conservation Germany) and the political activities with the SPD during the week also result in full schedule. That's why Gudrun often has to use Saturdays for her work, when Niko can take over the children . In the end, they only have Sunday for a joint family program. Gudrun knows that Niko is not happy with this arrangement. As a working wife, Gudrun cannot and does not want to wait for Niko during the week. The weekend together usually ends around midnight on Sundays. Niko then drives back to Hohenheim on empty highways, just about 250 km or three hours.

Three weeks later, on the way home from the Institute, Gudrun takes the newspaper and the mail from the mailbox. An airmail letter to Niko from Canada! Gudrun's heart jumps! She doesn't usually open letters addressed to Niko. That's, however, different now. Gudrun is sure that Niko also wants to know immediately. She runs up the stairs to the living room. She grasps the letter opener, Sellaperuma's farewell gift, from her desk and opens the envelope. She holds her breath and reads: "An invitation to Dr. Nikolaus Koeniger to give an application lecture at the University of Guelph at the beginning of December. The university in Guelph will cover the cost of the flight." Then many bureaucratic twists and turns, which are not so important for the time being! Gudrun hurries to the phone and calls the Institute in Hohenheim: "This is Mrs. Nolte, secretary of beekeeping, Hohenheim. What can I do for you?"

"I would like to speak to Dr. Niko Koeniger. It is urgent. I am his wife!"

"Wait, I'll put you through!"

Then comes a ringtone and after a while Mrs. Nolte again: "Dr. Koeniger does not answer. He is probably with the bees in the field. We can't reach him. Should I give him your message when he comes back?" This news from Canada does not have to make the rounds in Hohenheim before Niko is informed. "Please tell my husband to call me as soon as possible," Gudrun asks: "I have an important message for him!" "Won't you tell me the news for your husband? It's faster than a callback."

Mrs. Nolte is curious. All the better that she didn't say anything about Canada.

"Very kind of you, Mrs. Nolte, but it is about private matters that I have to discuss directly with my husband. Please just tell him to call me back as soon as possible! Thank you very much!"

Gudrun hangs up the phone and puts on the tea water. Yes, a cup of Sri Lankan tea, she needs it now. She realizes that this is only a first hurdle they have jumped successfully. She has experienced that Niko is usually successful in such situations. She wants to keep that to herself. Niko usually tends to be exaggeratedly optimistic. It's better to have subdued confidence, rather than water in the wine, that's the role Gudrun decides to take on. Will Niko notice that? After more than ten years of marriage, they know each other almost too well. Niko is safely back at the Institute in Hohenheim by noon, and he has not yet reported. Has her message not reached Niko? Resolutely, she chooses the extension of Professor Vorwohl, the acting director of the Institute, whom she met in Sri Lanka: "Hello Günther, this is Gudrun. Can you help and put me through to Niko?" "No problem, I'm happy to do that. Niko, as far as I have seen, is in his laboratory!" Then Niko on the line: "What's the matter, Gudrun? You certainly don't call without reason. Hopefully no misfortune?"

"No, Niko, a great stroke of luck. Didn't my request for a callback reach you? I spoke to Mrs. Nolte on the phone." "This secretary is impossible. What happened?"

"There is an invitation for you to give an application lecture in Guelph. On December seventh!" "Gudrun, that's great. A winter adventure in Canada. We should take a week. Arrive at least two, preferably three days before the lecture, so that my switch to English works. We get to know our future workplace and the city. Can you ask your mother to look after Anna and Fritz?" "The university in Guelph only pays for your flight and the invitation is only for you!" "Gudrun, we both have to go. Let us talk about it calmly. I take a vacation and leave right away. Tonight, I will be with you in Oberursel!" "Yes, I'm happy. Drive slowly!"

Gudrun immediately calls her mother in Wilhelmshaven.

"This is a great opportunity! And Niko is right. You absolutely have to fly with him. If you don't have money for your flight, we can help. I just want to talk to Aunt Ursel on the phone. She wanted to come here anyway and she can take care of dad while I look after your children in Oberursel. Give me a little time. So, from 4 to 14 December, grandchildren in Oberursel! I'll get back to you tonight and then I'll have everything under control!" Gudrun is speechless about her mother's generous offer. Obviously, the obstacles to the "joint winter adventure" in Canada have been overcome?

Forty-second chapter

Gudrun and Niko visit Guelph in Canada

On the approach to Toronto, the plane suddenly sinks, only to catch itself again just as suddenly with a jerk. Several times. Niko looks pale. Then the plane touches down. "Niko, come on, we've landed. We have to get out." Gudrun looks sorrowfully at Niko, who is still sitting in his seat with his eyes closed and pale. "Just a moment. I'm feeling better!" Niko walks slowly behind Gudrun to the exit. Gudrun finds the conveyor belt with the luggage from Frankfurt. Niko sits on the trolley while Gudrun retrieves the suitcase.

"What do you want in Canada at this time?" asks the officer at the immigration counter. "We are tourists. Canada in winter. We want to experience that," answers Gudrun. "Where do you want to go and where are you staying?" "We want to go to Guelph first. Holiday Inn." The official is satisfied to be able to classify them clearly and unambiguously in the drawer of "tourists". No further questions. After less than five minutes, visas for three months are stamped in their passports. They go through the green exit "Nothing to declare!". Gudrun takes out the last issue of "Bee World" and holds the magazine demonstratively in her hand as agreed. Professor McEwen does not show up. He had promised to pick them up.

"Gudrun, what should we do? We've been waiting here for more than an hour."

Niko obviously still doesn't feel well. "My suggestion: You wait here with the Bee World and I'll go, exchange money and buy Coke and sandwiches! If McEwen comes, you'll have to wait for me. I will hurry." When she returns, Niko eats three loaves of bread with ravenous appetite. He had skipped the last meals on the plane because of the nausea that was developing. "Thank you very much Gudrun, that was a great idea. I'm slowly getting full and I'm feeling better. Now Professor McEwen could come!" An older gentleman hurries up to her: "So sorry, I'm late? I'm McEwen and you're Dr. Gudrun and Dr. Niko Koeniger? Have you been waiting for a long time?" "Yes, immigration and customs were done in less than 10 minutes," Gudrun answers angrily and undiplomatically. "So fast? The last candidate took more than two hours. He explained that we would pay for his flight and the stay. The officer wanted to issue him a business visa. The necessary papers were missing. Finally, I had to explain to the official that we had invited him from the University of Guelph to an interview. He then got a tourist visa. What was that like for you? Didn't you tell people what you came here for?" "Of course! We want to see Canada in winter, stay at the Holiday Inn and are tourists. That's the truth!" Gudrun laughs and her laughter infects McEwen. "That's right and you have saved me trouble!

Now to my car. Unfortunately, the parking lot is further away."

Outside, they are greeted by a cutting, cold wind. Canada in winter. The drive to Guelph doesn't take long. It is pitch dark when McEwen drops them off in front of the hotel. He comes in and explains to the woman from the reception that they are guests of the university. A short reminder that Niko will be picked up the next day at 9 a.m., and "Good Bye".

Niko is tired. He doesn't want to go to the restaurant to eat. Gudrun finds the room card. "What do you want?" "Is there a burger with French fries?" Gudrun orders the burger for Niko and a Greek salad for herself. In addition, two cans of Budweiser. Room service comes quickly. One shower and both are in bed and sleeping.

The next two days, Gudrun and Niko tour the University of Guelph. They visit their colleagues and are talking about possible research projects. Professor R.W. Shuel, who has published important experimental work on bee sperm, welcomes Niko and Gudrun happily. It turns out that he has obviously studied Gudrun's publications very carefully. He enters directly into a discussion with Gudrun for a future cooperation on sperm storage. Professor Shuel suggests organizing a long-term work program based on Gudrun's queen bee surgical technique and applying for a grant for Gudrun. "Now you just have to successfully complete this application. Then your wife also has a job in Guelph." Gudrun is in high spirits.

The official meeting with colleagues in the Department of Environmental Biology, to which the advertised position belongs, takes place on a formal level. Gudrun is not allowed to participate. Niko has to answer questions about his teaching program and his research priorities. The expectations of him are formulated: The successful applicant should use half of his activities for development projects in Asia, Africa and later also South America for the Canadian development agency CIDA and the other half for the establishment of a research group in Guelph. "These are two jobs for which you need two people," says Niko. This remark is contradicted from various sides and a short discussion develops, in which Niko does not participate. The meeting ends in a rather tense atmosphere.

He meets Gudrun in the hotel: "That was a failure. I was stupid to say that you can't be in Sri Lanka and Guelph at the same time. I certainly won't get the job."

"Niko, you're crazy. The colleagues in the faculty will certainly reconsider your arguments. In addition, tomorrow afternoon there will be your lecture on the defensive behavior of tropical honey bees with the pictures from Sri Lanka and Pakistan. This will convince even the largest skeptics. I have the feeling that we are welcome here. You'll get the job." Gudrun radiates an enthusiasm that Niko does not share. He doesn't want to spoil her positive mood: "Gudrun, I'm very hungry. Let's go down to the restaurant."

Later, they sit contentedly over a sumptuous dinner and treat themselves to a bottle of California red wine.

The applause after Niko's lecture lasts surprisingly long. Nevertheless, Niko is only reassured when he sees the satisfied expression of Gudrun, who is usually one of his most critical listeners. Many questions and the discussion also reflect agreement, almost exuberance. Only towards the end does a critical voice ask how Niko intends to improve his English? "I've been in beautiful Canada for three days now and I'm hearing your perfect

English. Believe me, my English has gotten a little better every day this way. My answer is: I want to continue to listen to you carefully."

Some people laugh and applaud. Obviously—according to Gudrun's comment—Niko's English was not so bad!

The next day, Gudrun and Niko are invited to meet with Professor Gordon Townsend, who picks them up at the hotel. He had visited them in Sri Lanka and the two students from Peradeniya who worked with Niko were hired by him as bee consultants for his development project. The car stops at an old, lavishly restored farmhouse, which is located on a hill in the middle of a park-like area. As they get out, an elderly lady comes towards them: "This is Heather, my wife, meet Gudrun and Niko. Heather, you'll remember. I told you about the two of them, their children and their many students in Sri Lanka!"

"Welcome! I am happy to welcome you here. Gordon was very impressed with your household in Anuradhapura. So many children and students from Sri Lanka and from Germany. That must have been very nice?"

"Niko and I are also glad to meet you, Heather—may I call you Heather? Yes, the mixture of Sri Lanka and Germany was very productive and did us all good."

Gudrun wants to continue talking, but is interrupted: "Come in first. We don't have to endure the cold and wind any longer. It's warm inside!"

Heather opens the front door. They hang their coats on the coat rack and go into a large living room, in the middle of which stands a huge table decorated with many carvings. Space for at least twelve people.

"An impressive table? Probably not from Canada?" asks Niko.

"The table comes from East Africa, by Isaak Kigatira. You met him at the congress in Delhi," Gordon answers. "Probably a thank you for your project work in Kenya and for the development of the topbar hive. Truly a crucial simplification and improvement of beekeeping in the tropics!" "Bees again," Heather groans. "I see that it is no different with Niko than with Gordon? Gudrun, how did we deserve this. Please take a seat."

Heather points to the seating area. Then she brings a large teapot and fine porcelain from China or Japan. Gordon comes from the other side with four glasses.

"Sri Lankan tea or brandy? Unfortunately, I don't have coconut arrack, like you in Anuradh." "For me, first tea!" Gudrun wants to make a good impression.

"Gordon, a little brandy for me. My lecture is over and I want to enjoy the last two days in Guelph!" Gordon also takes a brandy. He raises his glass: "Cheers! Once again, welcome to Guelph and to your bright future in Canada!"

Gordon wants to show Niko his trees and their large garden. "Just go out in the cold. Gudrun and I stay here. Gudrun, if you feel like it, I'll show you my greenhouse, it's warmer there." Heather and Gudrun get up and walk off together. Gordon and Niko finish drinking. "Niko, another little sip?"

Gordon explains to Niko when and how he has planted the trees. Gordon is proud to explain the idea and the design. Niko feels the second glass of brandy and bravely asks for more details. Until he gets too cold. "Gordon, my clothes are more adapted to the climate of Sri Lanka than to Canada in winter. Can we go into the house?" "Sorry, Niko, very rude of me. Yes, let's get in quickly. Let's see what our ladies do."

Dinner is provided by a catering service. Gudrun and Heather unpack in the kitchen, while Gordon and Niko arrange the many different dishes on the large table in the living room. A hot Mulligatawny soup, unfortunately defused Canadian, with some samosas are the entrée. It continues Canadian with baked potatoes and roast turkey. Numerous very delicate salads and pickles complement the main course. Finally, a crispy baked apple from their own harvest. The conversation moves from Gudrun's and Niko's children to Heather and Gordon's grandchildren, it's about school and education. Then a taxi drives up. Gordon and Heather say goodbye to Gudrun and Niko and thank them warmly for the visit. Gudrun answers "That was a very nice evening with you. Your house, the park, flowers, all really impressive."

As soon as they sit in the taxi, Gudrun inquires: "What did Gordon say about your application and the selection process?". "Nothing at all. As a disciplined Canadian, he adheres exactly to the prescribed confidentiality. Not a word about my application. Only bragging about his farm and his impressive trees. Boring and cold. I was freezing miserably." Gudrun looks at Niko in amazement: "It was really different with Heather, that is to say, very interesting. Let me report when we get back to the hotel."

Over Californian red wine, Gudrun says: "Do you know that more than 100 serious applications have been received for the advertised Professorship? Many foreigners, Indians, Europeans, English and US colleagues. You were one of seven Germans! Unfortunately, Heather didn't know any names. Gordon told Heather that your presentation was very good and some committee members who initially had reservations about you will now support your application". "Did Heather also say how many applicants were invited to the interview and lecture?". "Yes, probably no less than five colleagues with you. You are number three. You owe the first success to me," Gudrun continues. "How is that?" "Heather laughed and said that the boss, Professor McEwen, was thrilled about our smooth entry into Canada." "Tourist and Holiday Inn", our formula for international borders, has made the rounds of the faculty." "Gudrun, Holiday Inn, that's not the most important point. With your accompaniment to Guelph, it is clear to everyone that I really want the job. Who would bring his wife at his own expense if he only applied to play poker at home for better conditions? Without you I wouldn't have had a chance here. Whether I will be successful with your help, we will have to wait and see! After all, I'm only number three." At first, Gudrun doesn't know what to say. "Niko, you're crazy again. You are the main person and I am only your wife here. Anyway, I'm terribly tired, let's go to the room!"

The following night they fly back to Frankfurt. Gudrun sits awake for a long time and looks at the sleeping Niko. Many thoughts and memories of their years together go through her head. Their first flight together, from Israel to Frankfurt in 1967, more than ten years earlier, comes to mind. Their two children. Her crisis as a housewife-only. She thinks of the slow and arduous path back to research. The wonderful joint work with Niko and the children in Sri Lanka. Much better than Gudrun had ever dreamed of. Now Niko's work in Hohenheim. Hundreds of kilometers away from her and the children! That won't be good in the long run, Gudrun is aware of that. She also doesn't know whether she will be able to cope with the abandonment of her projects in Frankfurt and the move to Hohenheim. This Professorship in Guelph would offer a perfect way out. The possible cooperation with Professor Shuel would also offer her an opportunity for research. How realistic is this dream? Gudrun moans and then, overwhelmed by tiredness, she falls asleep in the seat next to Niko.

The children are at school and Gudrun's mother is happy to meet them when they unlock the apartment door. "I've been waiting. Gudrun, Dad wants me to return as soon as possible. Tell me first. What was Canada like? Niko, will you get the Professorship?" "Mom, slow down! Canada was beautiful and Guelph is a cozy little town. Niko sold himself well. There are still a total of five candidates in the race. It will certainly take longer before we hear whether or not. How were Anna and Fritz?"

"Great!" Gudrun's mom beams: "We had our fun. Anna is very reasonable and helpful. Fritz, on the other hand, often has a mind of his own. You really have two wonderful children. Nevertheless, I would like to return to Wilhelmshaven today. There is a train at 2 p.m. Do you agree?"

"Thank you very much for your help," says Niko. "It was important that Gudrun came along. So, it was clear to everyone that we really wanted to come. I'll take you to the train station!" Niko carries their suitcase into the bedroom. Mummy's ready-packed travel bag is behind the door.

Forty-third chapter

Gudrun and Niko clear up misunderstandings

Then the phone rings. Mrs. Nolte, the secretary from Hohenheim, calls. Is Niko back from Canada? "Didn't you recognize my voice?" asks Niko back.

Mrs. Nolte laughs: "Of course. It's just that Mrs. Lehmann from the Munich Beekeepers' Association has called several times and asks whether you will give the promised lecture on Sunday or whether you are still in Canada?"

"How does Mrs. Lehmann know about my trip to Canada? Did you tell her? Please give me Mrs. Lehmann's phone number. I want to talk to her directly!" Niko is unusually loud on the phone. "What's the matter, Niko?" asks Gudrun.

"I'm sorry, I have to go to Hohenheim immediately. I'll drop Mom at the train station and then drive off. I can't come this weekend. I have to give a beekeeping lecture. I agreed to do that six months ago and I can't back out now," Niko explains somewhat desperately. "You won't see Anna and Fritz again for ten days! Have you considered that? And go to Hohenheim right now? You haven't recovered from the flight and the time change yet. Oh, for you, work comes first, the family always ends up in last place!" complains Gudrun. "You know that's not true! Please let's not argue now. I have no choice. I have to go," with these words Niko takes Mom's travel bag and his trolley. He wants to hug Gudrun. She turns away and disappears into the bedroom. "Then just not," Niko grumbles and goes to the car with his mother-in-law.

"Niko, I don't want to interfere. Gudrun is right. That's not how it works. What are you thinking?" Gudrun's mother intervenes. "Sorry, I can't change that at the moment," Niko answers, "I have to go to Munich and prepare the lecture. I didn't know anything about the Canada trip when I said yes to Munich. Please don't interfere!" They are silent until the train station. Gudrun's mother is offended. When saying goodbye, Niko says curtly: "Thank you, goodbye and have a good trip!"

Gudrun waits for Niko's call in the evening. The telephone system at the Hohenheim Institute cannot be reached from the outside between 5 p.m. and 9 a.m. She realizes that no news is usually good news. By 7 p.m. however, she is worried. Is Niko offended by her accusations? He must know that she is waiting for his message. Finally, the telephone rings: "Sorry Gudrun, my call is late. You've been waiting. I fell asleep!" Niko's voice sounds surprisingly friendly, Gudrun thinks. "Better late than never, Karlsson-on-the-roof says! Was the trip good? Mom arrived safely, and Dad was at the train station. She sounded quite happy. How are you?" asks Gudrun a little anxiously. "Good, after two hours of laboratory sleep. I'm going to start doing my presentation now. The people of Munich want to hear something about *Varroa* mites. Let me come back to the quarrel started in Oberursel" Niko sighs, "I'm sorry. I really should have stayed longer and left at night. I had completely forgotten about Munich." "I thought you had deliberately concealed this from me. Sorry!" replies Gudrun. "I didn't react well. Sorry!" Gudrun obviously wants peace.

"Niko, the children and I need you. Finish your talk and come here as soon as possible," says Gudrun, longing. We have been together day and night for the last nine days in Canada. "I'll be with you on Wednesday or Thursday at the latest. Kiss Anna and Fritz from me. I'll get back to you tomorrow after the lecture."

Niko hangs up the receiver without waiting for Gudrun's answer. He now urgently needs to switch gears and pick the slides for Munich.

Gudrun has difficulty getting used to everyday life in Oberursel again. Due to the time change, she falls asleep late. The alarm clock at six in the morning wakes her out of her deep sleep. Anna and Fritz demand more attention, probably because she has been away for a while and Niko has left without seeing them. Fritz causes trouble when getting dressed in the morning.

It rains on Saturday. Gudrun goes shopping together with Anna and Fritz and then prepares the meal. Anna is an enthusiastic potato peeler. Fritz rubs the peeled potatoes through a sieve. They add two eggs, a little flour and a little salt to the grated potatoes. Anna then carefully empties the porridge into the hot pan with a large spoon and smooths it out. The children start to form figures with the potato dough. Anna wants to fry a "potato fish" in the pan. Fritz has set himself a more complicated task: his potato pancakes should look like bears. Gudrun's attention is required: children and hot frying fat are a dangerous combination! At the end, there is a bowl of potato artwork on the table. The fish, the teddy bears and also the normal pancakes taste excellent with applesauce!

On Sunday, Niko calls late: "I've just returned to Hohenheim. Munich was quite good. Many questions and discussions. How are you and the children?" "We cooked potato pancakes together. Anna made a potato fish for you and Fritz a potato bear. When are you coming?" "On Wednesday afternoon there will be a team meeting here. I will try to drive off afterwards. Whether that will work is not certain. It could be that I am still needed here. Our *Varroa* mite experiment did not go well. Too many colonies were dead when we flew to Canada. If the mortality continues, we will have to stop the experiment. I won't know until tomorrow. I'm calling you!" "I haven't been in biochemistry yet and don't know what my program for the next few weeks will look like. I need to talk to Professor Fasold." Gudrun takes a break. "Do you think the application in Guelph will be successful?" "I'm not thinking about Canada right now. Today I have done the lecture in Munich, tomorrow it will be the *Varroa* mite experiment and the other programs. I have to submit the abstract for the Bangalore Congress and prepare the DFG proposal for travel expenses. You're coming to India with me, aren't you?" "Niko, you are incorrigible, your wife Gudrun has two children who need their mother because the father is not here. The bad thing is that of course I want to go to Bangalore!" Gudrun becomes quieter and quieter, then a sigh. "Dear Mother Gudrun. There are still three months to go until Bangalore. Time enough to find a good solution for all our problems. And the two of us are flying to India. Anna and Fritz are then looking forward to grandma and grandpa in Wilhelmshaven." "Niko, I have to hang up and put the children to bed."

When the children are asleep, Gudrun sits down in her armchair and thinks. She should take more care of the children. She can't tell exciting, fictitious stories like Niko in the evening. Even romping together in bed or adventures in the forest would not take place without him. Niko can't really assess that. The weekdays without him gradually go beyond Gudrun's strength. It needs a change. This change must come from her. Niko is responsible for the economic security of the family. This requires a high level of work in research. Gudrun knows that. She will certainly not be able to go to India with Niko. She has just been to Canada. There are all-day schools in Canada, she has learned, and Professor Shuel has expressed interest in her cooperation. He even knew the results of her doctoral thesis, although Gudrun had to publish the publication in German because of Professor Ruttner. The question of how the queen keeps the sperm alive for many years has still not been answered. And in the last ten years, the analytical methods of chemistry have been significantly refined. Now she could move forward.

In the course of the coming weeks, the children and she get used to Niko's absence during the week again. Anna and Fritz are now big enough to talk on the phone. This allows Gudrun to take part in BUND meetings, lectures and other events in the evenings. Mrs. Homann, her neighbor and friend, can be reached by phone and can help if they call. The weekend with Niko runs separately. Gudrun tries to prevent her other activities from spilling over into her time together with Niko. Niko does not miss the fact that a large part of family life takes place without him. Anna and Fritz are happy when he comes. More and more, however, they arrange to meet friends on Saturdays and disappear after lunch. Often the only thing left for him is the "good night story". For Gudrun, too, the weekend often seems to be an annoying interruption of her numerous projects during the week. He finally realizes that he has to end the separation from his family.

In the evening, he explains to Gudrun: "I will find an apartment for us in Hohenheim. Hopefully I'll find something in the near future." Gudrun reacts irritated: "Niko, that's not so easy. The children go to school here. A change of school is only possible at the end of the school year. My research project is based in Oberursel. Have you forgotten that the application for Canada is still open?" Niko tries hard not to let his disappointment about Gudrun's negative answer show: "I need you and the children every day! I expect the rejection from Guelph in the next few weeks. Until then, I want to use the time to look for a suitable place to stay in Hohenheim for us. With family only on weekends, it can't go on like this." "Oh Niko, we all feel the same way. We also miss you on a daily basis. Don't you want to wait at least until the decision from Canada?" "No and yes," Niko answers. "No, I don't want to wait idly any longer, I have to look for an apartment or a house for us now. Yes, we should not move until the decision from Canada is available." "It's not clear how that's supposed to work. With good housing offers, you have to decide quickly. Be that as it may: You can search. Only after the news from Canada will we talk about a relocation date!" Gudrun leaves no doubt that this decision is final to her.

Niko leaves the living room. Gudrun's clear and factual announcement hurts him. Worse still, he knows that Gudrun is right. In the case of good housing offers, an immediate commitment is required. He had expected more understanding and less objectivity from Gudrun. Gudrun senses Niko' disappointment. The next weekend, Niko reports on his search: "The real estate market in Hohenheim and the Stuttgart area is similarly narrow and expensive as here in the Rhine-Main area. We will certainly have to reckon with a higher rent. There is no alternative." "Well, there are always alternatives! A good, productive biologist like you can probably find a job here. Have you ever thought about coming back to Oberursel?" "Are you serious? That would mean that I would have to give up bee research!" "Don't snap right away. I know you don't want to give up the bees. And I can understand that well. Maybe you're right, a Niko without bee research is not an alternative." Gudrun takes him in her arms: "My dear Niko, our agreement stands. As soon as Canada is decided, we will come to Hohenheim or we will emigrate to Canada together. That would be by far the very best solution, wouldn't it?" Gudrun laughs. Niko is not in the mood to laugh.

"Now I understand. Bee researcher Gudrun would rather go to Canada than to Hohenheim. As long as you take me with you, both should be fine with me!" comments Niko. "Niko, you're crazy. It's not about Gudrun or Niko, but always about both bee researchers together. Together!" Gudrun clarifies. Finally, Niko smiles too.

Forty-fourth chapter

Gudrun, Niko and the children emigrate to Canada

On the way home from the Institute at noon Gudren passes by the mailbox. She takes a short break there, takes a deep breath and pushes the letterbox key into the lock. Between the Frankfurter Rundschau, the advertising and the other mail: no airmail letter from Canada! Over the course of the weeks, her tension should actually subside. Perhaps the discussions with Niko or the conversations with Anna and Fritz, with whom she talks about Canada, are responsible for the fact that the wait for this letter is just not easy. She doesn't react sensibly, does she? When Gudrun unlocks the apartment door, the phone rings. Probably one of her friends. She takes off her coat and goes into the kitchen. The phone can wait. When Gudrun picks up the phone, the conversation is no longer on the line. Well, she thinks, that was probably Sabine Wolf, she will certainly call again. Yes, after a few minutes the phone rings again. Gudrun picks up the phone and hears a deep voice asking in English who is on the line. "This is Dr. Gudrun Koeniger speaking. Who am I talking to?" "This is Professor McEwen of the University of Guelph. You remember?" "Of course, we visited some time ago." "Please, I have to speak to your husband."

"My husband works at the Institute in Hohenheim. You have called our house. I can't reach my husband until tonight." "I need a quick commitment from your husband. Can you give a binding commitment that Dr. Nikolaus Koeniger will accept our position? You know the conditions." "Yes, Professor McEwen, we, that is, my husband will take up the position and come to Guelph. This is binding, you can rely on it! My husband and I are looking forward to Guelph." "Thank you very much, Dr. Koeniger, this is good news. I am sending an official letter with the offer today. He should please return the enclosed forms to us signed." "It's okay. Thank you for your call!" "Good bye."

The doorbell rings. The children are coming back from school. "Mother, what's there to eat? I'm very hungry!" Anna kicks her heavy satchel into the corner and runs into the kitchen. "The meal is not ready yet. There are sausages, mashed potatoes and vegetables. It takes time until everything is warm. Do you want cookies or chocolate first for the bad hunger? Today is a special day!" "Why?" The children look at her expectantly. "I just got a call from Canada. Daddy got a job there. We will emigrate to Canada!" "I don't want to go to Canada. I want to stay here," Anna shouts resolutely. "Fritz, do you want to go to Canada?" "Of course. There are bears there and I want to see them. The wilderness with the bears in Canada, that's great. I want to go there!" "First of all, there are cookies and chocolate here. We are not going to Canada for a long time. For now, everything remains as always. Do you want to drink a Coke? Today is a good day, you can even get Coke! Now I have to warm up the lunch." Gudrun puts the pan on the stove while the children pounce on the sweets. They run away with the loot, each to their room. Then she remembers that she has to call Niko.

Mrs. Nolte connects her with Niko, who is on the phone right away: "How nice that you're calling. What's the matter, Gudrun?" "McEwen just called. You've got the job! They want you as soon as possible," she almost screams into the phone. "Gudrun, that's not a joke, is it? I had given up hope. I feel completely different. Gudrun, I'll call you back!" He hangs up. Niko speechless? She has not experienced that in all these years. The situation must have weighed heavily on him. These men! Always just pull yourself together. Under no circumstances show what you feel. Something smells burnt. The sausages! Quickly to the kitchen. Well, she can scrape off the black spots. "Anna, Fritz, the meal is ready." The half sausage on the plate, a small spoonful of mashed potatoes and the vegetables: the children poke around in them. Gudrun swallows. It's your own fault! After chocolate and biscuits, the food stays on the plate. What can I do? "What do you think of ice cream?" Whether ice cream or sausage with mashed potatoes doesn't matter today. The main thing is to eat together at the lunch table.

It is not until around five o'clock that the phone rings again: "Niko, why are you only calling now?" "We had an Institute meeting. It was about my research program for the coming year. I was able to initiate many projects. Now I have a guilty conscience. Did you firmly promise McEwen that we would come to Guelph? As soon as possible?"

"Yes, I promised in your name that you would come. McEwen was quite relieved and thanked me several times. It was good that I was in Guelph and that McEwen knew me. He had no doubt about my acceptance. Niko, tell me, when are you coming? I put a bottle in the fridge!" "Gudrun, there's so much going on in my head. We have to keep a cool head despite all the joy." "Niko, you're crazy! This happiness! And your only concern is a cool head! Sometimes I ask myself, what happened to the great Niko I met in Freiburg many years ago? You get in the car today and come here immediately!" "Gudrun, I love you. Now run to the car. See you soon."

It is late when Niko arrives in Oberursel. The children are asleep. Gudrun stands in the apartment door in her old Bedouin dress from Israel, beaming with joy.

"Here I am." Niko jumps up the last steps. "Finally!" is all Gudrun can say when Niko takes her in his arms and carries her into the apartment.

The University of Guelph must submit Niko' documents to the Canadian Immigration Service. There it is checked whether there is no Canadian applicant who also meets the recruitment requirements for this honey bee Professorship. Six weeks pass, without news. Niko is worried. He announced in Hohenheim that he will take up a Professorship in Canada and will quit his employment at the end of the year.

"What if the authorities find a Canadian applicant after all and we don't get an immigration permit?"

At some point, Gudrun picks up the phone and calls Professor McEwen

"Hello Professor McEwen, this is Dr. Gudrun Koeniger. Do you have any news from the immigration authorities?" "Yes, I spoke to the clerk on the phone a few days ago. Everything looks good. There are no complications. They have to wait for the okay from Ottawa. I expect the decision in the next few days. Then you have the status of "Landed Immigrant" and can come. We expect you to arrive here in July and that your husband will give the first lectures and courses in the coming winter semester. That is the state of affairs. Let me add: We have brought in some colleagues from abroad and so far, it has never gone as smoothly as it did for you!"

Gudrun immediately calls Hohenheim: "Niko, I have just spoken to Professor McEwen. Everything is great. They firmly expect us to be there by July at the latest. Stop worrying. So far, we've done everything right. Should I list: Israel, the doctoral thesis, Pakistan, Sri Lanka ..."

"Stop it, Gudrun. I don't know why these doubts plague me. Is this perhaps due to Hohenheim? A state Institute for beekeeping is not a university Institute. Gudrun, it's frustrating. My manuscript on "Food Competition" is finally finished. I gave it to the secretary for a final copy. After four days, it was only halfway done. This woman has made many mistakes. Unimaginable. I'll have to type the manuscript myself."

"Niko, don't you have any other worries? I'm working my legs out here to organize our start to Canada and what are you doing? You quarrel with a typist. This won't do. You are needed here now!" "I still have to work in Hohenheim! I can't change that!"

"You have to change that!" The conversation becomes loud.

"Go to the administration and ask for vacation. If necessary, even unpaid leave. We will fly to Guelph by the end of July at the latest. Only three weeks left. And by then, everything here must be empty. I can't do it alone," Gudrun complains.

"Do you have the tickets? Do you know when exactly we are flying?" asks Niko. "On July 29 at 7 p.m. with Air Canada to Toronto. I arranged that with McEwen yesterday. The flight tickets will be sent to us next week. Understand, your place is now here with the family," Gudrun complains. "Gudrun, just few more days in Hohenheim. I think on Tuesday or Wednesday I will finally be with you and the children. I promise!"

When Niko arrives in Oberursel as promised, she is relieved: "It's good that you're finally here. You have to find the solution to squaring the circle. We have to sell and get rid of everything that is in the apartment. On the morning of July 29, the apartment must be empty. The new tenant, Mr. Rasel, wants to move in on July 30. Nevertheless, we must be able to live here until the 29th." Niko calls Mr. Rasel and explains that there has been an unforeseen postponement of the date and that the earliest move-in date is probably August 15. "We will not be able to confirm this date for a few days!"

Gudrun puts her own trolley in the children's room. "You can pack the things you really want to take with you. We can only take a little, only the most important things with us. That's difficult for all of us."

Anna has great difficulties: "Mother, what happens to all the things I can't take with me? What will happen to my beautiful closet from grandma, in which all my clothes and also the books are stored? You said that we will stay in Canada and I can't take my closet with me, can I? And neither does my bed," she is close to tears.

"You're right. Daddy and I can't pack something like that for us either. But in Canada you will get a new beautiful closet. You will go with Daddy to a shop with lots of cupboards and choose the most beautiful one! You can also choose a new bed yourself! If you've slept in the new bed many times, I'm sure you'll like it as much as this one."

Anna sighs and turns to the trolley in which she is supposed to store her hand luggage. She packs it full of her books. When Gudrun comes back in, she sees that it weighs well over ten kilos. How can she avoid a conflict? She puts Anna's favorite things, the cuddly blanket, the cozy pyjamas, the warm underwear, two sweaters and the socks from grandma on Anna's bed: "Don't you want to think again about what you need in Canada? It can be very cold there. I've put some warm clothes on your bed here!"

Later, Gudrun takes a look. Yes, her tactics worked. Anna has packed her clothes and her favorite blanket. There are only three books left, Red Zora, Karlson-of-the- Roof and Dr. Doolittle.

Fritz, on the other hand, wants to pack all seven of his strong bears into the trolley. "Mother, I need your big blue suitcase. My bears don't fit in this trolley."

"You know, we're flying to Canada by plane and we can only take a few things with us. Can't you choose the strong bear you like best? We will send the other six to grandma and grandpa in Wilhelmshaven. You can visit them there!" "No! I love all my bears the same! Everyone has to come along, otherwise I'll stay here with my bears!"

"Fritz, taking seven bears with you, that's really not possible. Just ask Daddy."

Niko says: "Fritz, I can understand you. If you don't come with me because you're staying with the bears, Mother and Anna will also stay here and I'll have to go alone without you. Canada is further away than Hohenheim. You will be without me. I'm very sad then, I'll miss you. Whatever happens, the family, you, Anna, Mother and I, we have to go to Canada together. I'm going to take one of your bears in my carry-on luggage. Then you have two bears. Maybe Mother will take one with her? And as I know Anna, she doesn't say no either. If it works, you will have four bears, the father bear, the mother bear and both bear cubs on the plane. Isn't that enough?"

"No, that's not possible," Fritz answers meekly. "Then what about the other bears?"

"We will then have to pack the other three bears into the big suitcases. For this we will unpack things and leave them here that we could actually use there as well. Otherwise, there is no space in the suitcases. Why don't you ask Mother if she agrees?"

Fritz runs to Gudrun: "You know, your father wants to take all the strong bears with him!" He explains the plan. "Fritz, your father is crazy. Why do all seven bears have to come along?" "That's clear. Father bear, mother bear and both bear cubs sleep with me in my bed as always. The other three bears have to guard us."

"And your father agrees to that?" Fritz gives Gudrun a bear cub: "For your hand luggage!"

Gudrun and Niko push the loaded baggage carts to the Air Canada counter. Niko heaves the suitcases onto the baggage belt. Gudrun hands over the passports and flight passes to the stewardess and gets the boarding passes. They watch as the last of their large suitcases starts on the conveyor belt and disappear. There are still almost two hours until check-in. Gudrun looks at Anna and Fritz: "Finally, that's done. I'm hungry now." Anna is also hungry. Fritz is thirsty: "I want something to drink. There's Coke here, isn't it? And I want to eat too."

"A Big Mac, fries with mayo and ketchup, that sounds very tempting," Niko notes.

After dinner, the children get restless: "We can't miss the plane and I don't want to run, like I did in Sri Lanka," Anna recalls. "I still have to go to the toilet!"

Niko gets up. "Let's clear the table and go!"

When the children and Gudrun come out of the toilets, it is really time.

"Where are you? We have to get to our flight."

They hurry with quick steps to Gate D32. You have two boarding tickets for the right side and two for the left side. The stewardess wants to separate the family. "Only two people to the right!"

Fritz protests: "We belong together!" Gudrun explains the problem: "Fritz, we're sitting in the middle of the plane. This means that two of us are now going to the right and two of us are only in the next row to the right. We'll meet in row 11 and sit next to each other!"

As they sit, Fritz asks: "Daddy, why are we going to Canada? You always go somewhere else. It's good that we can all come along this time. Like Sri Lanka. That was nice." Anna says: "Fritz, we have the best parents. Always somewhere else. Always new adventures. And now to Canada. Lots of snow and ice and reindeers. I want a dog sled." "Yes, Anna, a dog sled, and we'll take it to the bears. I want to see the bears in the wild. Bears are the best animals," Fritz agrees.

"We can take care of dog sleds, reindeer and bears later. First of all, we need a house with a kitchen, children's rooms and furniture," Gudrun objects. "That's clear," Anna notes. "Like in Sri Lanka. First we have to settle in and then come the adventures." "For us, Canada is not like Sri Lanka, but more like Oberursel," Niko intervenes. "Gudrun and I will do research with bees there and work at the University. You'll go to school. We will settle in Canada. We're going to be Canadians."

"I don't want to become a Canadian," says Fritz indignantly. "I want to remain the Fritz that I am!" Gudrun leans over and takes him in her arms: "You will always be our dear Fritz. Anna will remain Anna, Daddy and I will also remain as we are. Canadian only means that we live in Canada." "Okay, then I want to become a Canadian," says Fritz now comforted and satisfied. "And see the bears." It doesn't take long and both children fall asleep.

"Oh Niko, a long way from our first joint zoological internship in Freiburg to here on the plane to Canada for your first professorship. Your determination and our consistent commitment to bee research have paid off." "This is not only due to bee research. Rather, it is the post-war period, Gudrun, that has carried us! With the lost war and the Nazi terror until 1945, our country and many traditions were destroyed. A new start and departure were necessary. We were needed! At the same time, we had the unique opportunity to do some things differently and, above all, to get to know other countries. In any case, everything was different from our parents. Many people of our generation have taken advantage of this, not just us bee researchers."

"Bee research or the post-war period, both should be fine with me, Niko. Be that as it may, the last time in Oberursel with the children without you was not good. I'm glad to be sitting here on the plane!" Niko agrees. Then he falls asleep, his head leaning on Gudrun's shoulder. Gudrun can't sleep like this. Should she push Niko's head back? Or would she rather not? After a while, the sleeping Niko turns to the other side. Gudrun casts a grateful glance at her husband before closing her eyes as well.

The next morning in Toronto, Gudrun, Niko, Anna and Fritz are "landed immigrants".

Epilogue 2025

"Niko, you're crazy! Come down at once!"

Gudrun stands under the large cherry tree and looks up at Niko, who is laboriously trying to reach the ripe cherries on a branch high above him and put them into a half-full bucket.

"You are over 80 years old at standing at a great height on thin branches—will you never become sensible? Come down immediately!"

Gudrun gets loud. The passers-by stop behind the fence and look up at him. It really seems to be time to end the cherry harvest:

"I'm coming. Can you hold on to the ladder?"

Niko is out of breath as he stands next to Gudrun and regretfully shows the bucket, which is only half-filled with cherries:

"Another half hour and the bucket would have been full."

"Or perhaps you could have crashed and I'd be on my way to the hospital with you!"

"Obviously, we still experience situations differently—and that after all these years!" comments Niko.

"Yes, more than 60 years together. Who would have expected that in 1962? I certainly didn't!" Gudrun leaves no doubts.

Niko looks at his wife and remembers the self-confident, enterprising and critical student:

"Yes Gudrun, that wasn't love at first sight," Niko laughs.

"Not even the second one," Gudrun replies, "At that time we both only wanted to study and nothing else."

"And then the bees. The bees to this day!"

Niko points to the bee colonies under the cherry tree. "We stumbled through our lives together with bees," Niko continues.

"Why is that? I didn't stumble for 60 years," Gudrun replies indignantly.

"How did you experience our years together?" Niko looks at Gudrun expectantly.

"We always had a plan and a goal! Have you forgotten that? First the doctorate. We really wanted to become bee researchers. Stumbled far from it! That was consistent career planning. Do you want to deny that?"

Niko grins, he wants to provoke Gudrun: "I had planned exactly to meet the pretty student Gudrun from Wilhelmshaven in 1962 at the Zoological Institute of the University of Freiburg," Niko mocks and laughs.

"Given, Niko, of course there are unforeseeable coincidences. I don't want to deny that. You can only take advantage of opportunities and chances successfully if you know what you want to achieve."

Gudrun pauses for a moment and continues: "We always planned and always knew what we wanted! After the doctorate, the children. Then I wanted to go back to research. Bee research in Oberursel, Canada, Asia and Africa. Today, bee researchers work worldwide who learned their trade with us."

Gudrun looks at Niko expectantly: "Or do you see it differently?"

"No, that was probably the case." Niko sounds conciliatory: "I only have doubts as to whether everything went according to plan. Maybe we have mostly stumbled through the years with a lot of luck. In any case, the day we met by chance during our zoological internship was the most important day in my life. Also, in yours?"

Gudrun nods, hugs Niko and gives him a kiss!

Fig. 42. Spring 2025. Niko picks cherries

Fig. 43. The four species of *Apis* the Koenigers studied in this book. Clockwise from upper left: *Apis mellifera, A. florea, A. cerana, and A. dorsata.*